3

MONOGRAPHS ON
APPLIED PROBABILITY AND STATISTICS

General Editors

M. S. BARTLETT, F. R. S. *and* D. R. COX, F. R. S.

POINT PROCESSES

Point Processes

D. R. COX

Department of Mathematics,
Imperial College, London

VALERIE ISHAM

Department of Statistics and Computer Science,
University College London

LONDON AND NEW YORK

CHAPMAN AND HALL

First published 1980
by Chapman and Hall Ltd.
11 *New Fetter Lane, London EC4P 4EE*

Published in the USA by
Chapman and Hall
in association with Methuen, Inc.
733 *Third Avenue, New York, NY* 10017

© 1980 *D. R. Cox and V. Isham*

Printed in Great Britain at the
University Press, Cambridge

British Library Cataloguing in Publication Data
Cox, David Roxbee
  Point processes. —(Monographs on applied
  probability and statistics).
  1. Point processes
  I. Title II. Isham, Valerie III. Series
  519.2   QA274.42

ISBN 0-412-21910-7

Contents

Preface

Stochastic processes whose realizations consist of point events in time or space arise in many fields of application and have been extensively studied in recent years from several points of view. In this book we aim to describe recent work on the properties of such probabilistic models, putting emphasis on results and methods directly useful in applications. We treat neither the more abstract parts of the general theory nor the statistical analysis of data from point processes.

We presuppose a thorough working knowledge of elementary probability theory and, for some of the more specialized portions, some acquaintance with the simpler aspects of Markov processes is required. We have adopted as elementary a mathematical level as is reasonably possible, in the hope that the book will be useful both to students and research workers in probability and statistics and also to research workers in other fields wishing to apply stochastic point processes.

Some of the book was written while one of us (V. I.) held an S. R. C. Postdoctoral Research Fellowship, which is acknowledged with thanks. We are grateful to Dr C. J. Isham, Department of Physics, Imperial College for helpful comments.

London, April 1979   D. R. Cox
                                                        Valerie Isham

CHAPTER 1

Introduction

1.1 Preliminary remarks

This book deals with a particular kind of random process. The central idea is to study random collections of point occurrences. For the most part we consider the points as occurring along a time axis, although later we do allow other possibilities, for example that the points occur in some region of space.

A few outline examples illustrate the breadth of potential applications:

(i) emissions from a radioactive source occur in an irregular sequence in time, each emission defining a time instant;

(ii) Fig. 1.1 shows a small section of a time series of electrical energy in a nerve fibre. The occurrences of peaks define a sequence of points in time. If attention is concentrated on this sequence of time points, rather than on the magnitudes of the peak signals, we have a point process derived from a more complicated process;

(iii) Fig. 1.2 shows part of the sequence of dates of coal-mining disasters in Great Britain for 1851–1975, a disaster being defined as involving the death of 10 or more men;

(iv) in road traffic studies, we may consider the sequence of time points at which vehicles pass a reference point. Alternatively, if we examine a length of road at one fixed time instant and regard the position of each vehicle as specified by a point, e.g. by the position of its front wheels, we have a point process in one-dimensional space, rather than in time;

(v) many of the stochastic problems of operational research involve a point process. The instants of arrival of customers in a queue, the instants of withdrawal of items from a store, and the instants of failure of a component in some system are all examples;

(vi) in several of these examples, each point may be classified into one of several different classes or types. For example, in (iv), we may distinguish two classes, cars and lorries. In a queueing problem, (v), there may be several types of customer; alternatively, we may wish to

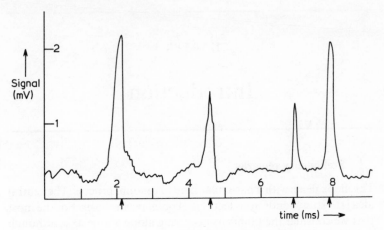

Fig. 1.1. *Electrical signal in nerve fibre.* ↑, *points of associated point process.*

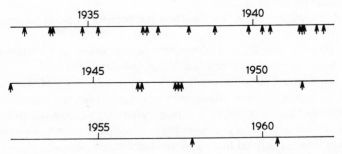

Fig. 1.2. *Coal mining disasters in U.K. killing* 10 *or more. Extract from larger set of data (Jarrett, 1979).* ↑, *times of disasters.*

consider both customer arrivals and customer departures as points, but as points of two different types. A point process like this in which several classes of point are distinguished is called multivariate;

(vii) point processes occurring along a single dimension in space rather than in time occur, for example, in plant ecology, when a line transect is taken in a field and the position along it of plants of a particular species is noted. Such applications give rise also in an obvious way to point processes in spaces of two or more dimensions.

There are three broad aspects to the theoretical study of point processes:

(a) the general theory, with particular stress on generality of formulation and on the exploration of conditions for existence and uniqueness of the various functions associated with the process;

(b) the study of special processes of potential importance in applications, including the derivation of techniques for investigating such processes assuming reasonably 'good' behaviour of all properties studied;

(c) the development of techniques for the statistical analysis of data from such processes.

The present book does not deal with (c). It concentrates mainly on (b), although, of course, the distinction between (a) and (b) is anything but clearcut. We shall, however, largely ignore questions connected with the precise formulation of regularity conditions and will rely extensively on informal arguments.

The borderline between point processes and a number of other kinds of stochastic process is not sharply defined. In particular, any stochastic process in continuous time in which the sample paths are step functions, and therefore any process with a discrete state space, is associated with a point process, where a point is a time of transition, or, more generally, a time of entry into a pre-assigned state or set of states. Whether it is useful to look at a particular process in this way depends on the purpose of the analysis.

1.2 Four illustrative examples

The properties of a variety of special processes important in applications will be studied in Chapter 3. In the present section, using a temporal approach, we outline properties of four special types of process, mainly in order to establish some basic ideas and notation, and also to prepare the way for the general discussion of Chapter 2.

(i) Poisson processes

The simplest point process is one in which points occur totally randomly in a sense now to be defined. Consider the process as defined over the whole time axis $(-\infty, \infty)$. Let $\mathcal{H}_t$ denote the history of the process at time t, i.e. a specification of the positions of all points in $(-\infty, t]$. For $u < v$, let $N(u, v)$ be a random variable giving the number of points in $(u, v]$. Then for a given positive constant ρ, with dimensions $[\text{time}]^{-1}$, the Poisson process of rate ρ is defined by the requirements that for all t, as $\delta \to 0+$,

$$\text{pr}\{N(t, t+\delta) = 1 \,|\, \mathcal{H}_t\} = \rho\delta + o(\delta), \tag{1.1}$$

$$\text{pr}\{N(t, t+\delta) > 1 \,|\, \mathcal{H}_t\} = o(\delta), \tag{1.2}$$

so that

$$\text{pr}\{N(t, t+\delta) = 0 \,|\, \mathcal{H}_t\} = 1 - \rho\delta + o(\delta). \tag{1.3}$$

An important aspect of (1.1)–(1.3) is that the probabilities concerned do not depend on $\mathscr{H}_t$, so that, in particular, the probability of finding a point in $(t, t + \delta]$ does not depend on whether there have been relatively few or relatively many points just before t, or indeed on whether there is a point exactly at t. It can be seen that any event defined by occurrences in (t, ∞) is independent of $\mathscr{H}_t$.

The requirement (1.2) virtually excludes the possibility of multiple simultaneous occurrences, i.e. of more than one point at the same time instant.

A final essential element in (1.1)–(1.3) is that ρ is constant, i.e. does not depend on t. For some purposes, however, for example to represent situations with a time trend or cyclical fluctuations in the rate of occurrence, it is useful to replace the constant ρ by a function of time, $\rho(t)$, say, the other assumptions remaining the same. This gives what is called a non-homogeneous Poisson process. A closely related possibility is that there is an observed explanatory variable, $z(t)$ say, and that the rate of the Poisson process at time t is a function of $z(t)$. For instance, in the example of Fig. 1.2 concerning coal-mining disasters, $z(t)$ may be the number of men employed at time t, or some other index of unit of risk at time t, and we may consider a model in which $\rho(t) = \rho z(t)$. In this book, however, by the unqualified term Poisson process we mean the process (1.1)–(1.3) with constant ρ.

Fig. 1.3(i) shows a typical short realization from a Poisson process.

The elementary properties of the Poisson process are well-known and will not be developed in detail here. Two key results are as follows.

First, let X denote the random variable representing the time interval from a fixed time origin, say $t = 0$, to the first subsequent point. Because the development of the process in the region $t > 0$ is independent of $\mathscr{H}_0$, it is immaterial whether there is a point at $t = 0$. Then the probability density function (p.d.f.), f_X, and survivor function, $\mathscr{F}_X$, of X are

$$f_X(x) = \rho e^{-\rho x}, \ \mathscr{F}_X(x) = \mathrm{pr}(X > x) = e^{-\rho x} \quad (x > 0), \qquad (1.4)$$

i.e. X is exponentially distributed with parameter ρ and mean $1/\rho$. One proof proceeds via the fact that for $x_1, x_2 > 0$

$$\begin{aligned} \mathscr{F}_X(x_1 + x_2) &= \mathrm{pr}(X > x_1)\mathrm{pr}(X > x_1 + x_2 | X > x_1) \\ &= \mathrm{pr}(X > x_1)\mathrm{pr}(X > x_2) \\ &= \mathscr{F}_X(x_1)\mathscr{F}_X(x_2), \end{aligned} \qquad (1.5)$$

the second line following because the condition $X > x_1$ implies that

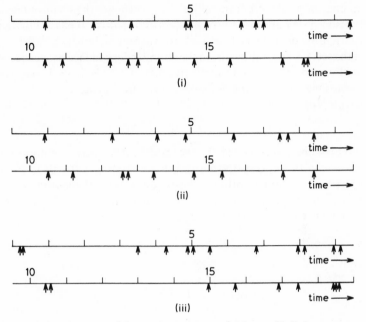

Fig. 1.3. *Realizations of three point processes of unit rate.* (i) *Poisson process.* (ii) *Renewal process, gamma distribution for intervals, coefficient of variation* $1/\sqrt{2}$. (iii) *Renewal process, gamma distribution for intervals, coefficient of variation* $\sqrt{2}$.

we can take a new origin at x_1. The functional equation (1.5), together with the initial condition $\mathscr{F}_X(0) = 1$, implies (1.4).

Further, the above definition of the Poisson process implies that, starting from an arbitrary time origin, subsequent points are at times

$$X_1, X_1 + X_2, X_1 + X_2 + X_3, \ldots, \qquad (1.6)$$

where the random variables $\{X_i\}$ are independent and identically distributed with the exponential distribution (1.4). This property provides an alternative definition of the Poisson process, and a convenient way of simulating it.

Note from (1.6) that the rth point after the origin occurs at time $T_r = X_1 + \ldots + X_r$ and that this has a gamma distribution, with density

$$\rho(\rho t)^{r-1} e^{-\rho t}/(r-1)! \,.$$

For the second key property of the Poisson process, consider the number, $N(a, b)$, of points in a fixed interval $(a, b]$; slightly more generally we may consider an arbitrary set A on the time axis and the

number of points $N(A)$ therein. It can be shown that $N(a, b)$ has a
Poisson distribution of mean $\rho(b - a)$ and that $N(A)$ has a Poisson
distribution of mean $\rho|A|$, where $|A|$ is the length (Lebesgue measure)
of the set A. For, it follows from the strong independence properties of
the Poisson process that the distribution of $N(a, b)$ depends only on
$b - a$. One proof of the Poisson distribution proceeds by writing
$N(t) = N(0, t)$ and

$$p_r(t) = \text{pr}\{N(t) = r\} \quad (r = 0, 1, \ldots),$$

and then obtaining and solving the forward differential equations for
the stochastic process $\{N(t)\}$, that is by relating $p_r(t + \delta)$ to $p_r(t)$,
$p_{r-1}(t), \ldots$. Another proof is based on a general relation which
applies to any process of single points: that for $r = 1, 2, \ldots$ the event
$N(t) < r$ is equivalent to the event

$$X_1 + \ldots + X_r > t. \tag{1.7}$$

Thus

$$\text{pr}\{N(t) < r\} = \int_t^\infty \frac{\rho(\rho u)^{r-1} e^{-\rho u}}{(r-1)!} du$$

$$= \sum_{s=0}^{r-1} \frac{e^{-\rho t}(\rho t)^s}{s!}, \tag{1.8}$$

on repeated integration by parts. The Poisson distribution for $N(t)$
follows directly from (1.8). The fact that $E\{N(t)\} = \rho t$ justifies the
term rate for the parameter ρ.

More generally the numbers of points in a collection of non-
overlapping intervals have independent Poisson distributions. The
distribution of $N(A)$ for an arbitrary set A follows on expressing A as a
union of a finite (or countably infinite) number of non-overlapping
intervals and applying the additive property of the Poisson distribution. More generally still, if $A_1, A_2, \ldots$ are arbitrary non-
overlapping sets, the random variables $N(A_1), N(A_2), \ldots$ have
independent Poisson distributions of means $\rho|A_1|, \rho|A_2|, \ldots$.

In fact, the Poisson process could be defined by any one of the
following mutually equivalent properties:

(a) by the original properties (1.1)–(1.2), which we call an intensity
specification;

(b) by the property that starting from the time origin, the intervals
$X_1, X_2, \ldots$ between successive points are independently exponentially distributed with parameter ρ, which we call an interval
specification;

(c) by the above specification of the joint distribution of $N(A_1)$, $N(A_2)$, ... for an arbitrary collection $A_1, A_2, \ldots$ of non-overlapping sets on the time axis, which we call a counting specification.

The interplay between these three kinds of specification is a recurring theme in the study of point processes. For the moment, note that (a) can be used for constructing a realization of a Poisson process, that (b) gives an extremely efficient basis for such construction, whereas (c) is not directly useful in this respect. If the points are distributed in a space of more than one dimension, the specification in terms of an analogue of the intervals is rather less useful.

We shall see later that the Poisson process plays a central role in most aspects of the theory of point processes.

(ii) *Renewal processes*

An important class of point processes, generalizing the Poisson process, is obtained from the interval specification (b) above by a fairly simple extension. Suppose that the intervals are independently distributed, but that instead of their density being necessarily of the exponential form (1.4), it is, say, g. More generally, to cover the possibility of discrete components, we may let the intervals have a distribution function G, although if G has an atom at zero then multiple simultaneous occurrences are possible; we exclude this possibility for the moment.

If the process starts from time $t = 0$, then a little care is needed with the initial conditions. If a point is known to have occurred at $t = 0$, then we define a process in which $\{X_1, X_2, \ldots\}$ are independently and identically distributed with density g and call this an ordinary renewal process. To allow the possibility of other initial conditions, suppose that X_1 has a density g_1 and that $\{X_2, X_3, \ldots\}$ are as before, all Xs being independent; we call this a modified renewal process. Now suppose that an ordinary renewal process starts with a point not at $t = 0$ but at $t = t^0$, where $t^0 \to -\infty$, and that we consider the development of the process from $t = 0$ onwards. It is clear that in general the time X_1 from $t = 0$ to the first subsequent event, called the forward recurrence time from $t = 0$, will not have the density g and that if we examine the process for $t > 0$ a modified renewal process will be obtained, but with a special form for g_1.

This special form can be calculated as follows. The chance that the origin falls in an interval of length in $(z, z + \delta)$ is proportional to $zg(z)\delta + o(\delta)$ and hence, on normalization, the density of the length of that interval is $zg(z)/\mu_X$, where $\mu_X = E(X)$ is the mean of the density g. This type of sampling is known as length-biased sampling. Given that the origin falls in an interval of length z, the time X_1 from the origin to the next point is uniformly distributed over $(0, z]$, so that finally the density of X_1 is

$$\int_x^\infty \frac{1}{z}\frac{zg(z)dz}{\mu_X} = \frac{\mathscr{G}(x)}{\mu_X}, \tag{1.9}$$

where $\mathscr{G}$ is the survivor function corresponding to the density g. The above argument uses the common distribution of the intervals but not their independence. Thus (1.9) applies more generally than just to renewal processes.

For our final definition, then, we specialize the modified renewal process by requiring the density of the first interval to be given by (1.9), thus obtaining what is called an equilibrium renewal process.

It can be shown directly from (1.9) that the equilibrium renewal process and the corresponding ordinary renewal process are identical if and only if g is exponential and the process is a Poisson process. Then the special consideration of initial conditions in defining the process on $(0, \infty)$ is unnecessary, as is clear from the strong independence assumptions occurring in the definition of the Poisson process.

The elementary properties of renewal processes are discussed in detail by Cox (1962). Many of the ideas that occur in the present book arise in simpler form in connection with renewal processes. One such general matter, that has already arisen above, distinguishes processes in which it is given that a point occurs at the time origin from those in which the time origin is arbitrary, i.e. in which the process started in the remote past. It is convenient to say that the process on $(0, \infty)$ is formed by synchronous sampling in the first case and by asynchronous sampling in the second; the sampling referred to is the effective choice of time origin.

In a general way, if the density g is relatively less dispersed than an exponential density, e.g. if it has coefficient of variation less than one, the point process is qualitatively more regular than a Poisson process. Correspondingly, if g is more dispersed than the exponential density, the point process is even more irregular and 'clustered' than a Poisson process. Fig. 1.3(ii) and (iii) show some simulated data, based on gamma forms for g.

While by suitable choice of the density g it is possible to produce renewal processes with a wide range of qualitative behaviour, the renewal process remains very special because of the strong requirement that the intervals between successive events are mutually independent.

(iii) *Linear self-exciting processes*

A recurring theme in this book is the importance of the complete intensity function $\rho(t; \mathcal{H}_t)$ defined by

$$\rho(t; \mathcal{H}_t) = \lim_{\delta \to 0+} \delta^{-1} \text{pr}\{N(t, t+\delta) > 0 | \mathcal{H}_t\}, \qquad (1.10)$$

where $\mathcal{H}_t$ specifies the point process up to and including t.

For a process satisfying

$$\text{pr}\{N(t, t+\delta) > 1\} = o(\delta) \qquad (1.11)$$

to be physically well defined, it is necessary that we should be able to find the probability of a point in $(t, t+\delta)$ given any realization $\mathcal{H}_t$, so that for such a process (1.10) is effectively uniquely defined and determines the probability structure of the point process. Condition (1.11) ensures that there are essentially no multiple simultaneous occurrences; this will be discussed further in Chapter 2. For the Poisson process, the complete intensity function is a constant, the rate ρ. For a renewal process, the independence of the intervals between successive events implies that (1.10) involves $\mathcal{H}_t$ only through the time from t back to the point, if any, immediately preceding t. It is natural to build up simple processes by supposing $\rho(t; \mathcal{H}_t)$ to involve $\mathcal{H}_t$ in specific ways.

One possibility is to suppose that each point in the past influences $\rho(t; \mathcal{H}_t)$ in a way decaying in time and that contributions from distinct points add together. Let $w(x)$ be a non-negative weight function defined for $x \geq 0$ and let points occur at times $\ldots, t_{-1}, t_0, t_1, \ldots$. Then the linear self-exciting process defined for all t has

$$\rho(t; \mathcal{H}_t) = \gamma + \sum_{t_i \leq t} w(t - t_i), \qquad (1.12)$$

where γ is a positive constant.

A notation in many ways more convenient than (1.12) is to write, by definition,

$$\rho(t; \mathcal{H}_t) = \gamma + \int_{-\infty}^{t} w(t - z) dN(z), \qquad (1.13)$$

where $\{N(z)\}$ is a stochastic process cumulating the number of points from a time origin in the remote past. In this notation the number of points in $(a, b]$, for $b > a$, is written

$$N(a, b) = \int_{a}^{b} dN(z).$$

Now if (1.13) represents a point process in which points occur at a constant rate ρ, in fact determined by $E\{\rho(t; \mathscr{H}_t)\}$, then, on taking expectations in (1.13), we have that

$$\rho = \gamma + \rho \int_0^\infty w(x)dx, \qquad (1.14)$$

which relates the rate ρ to the defining parameters γ and w. Clearly, in order that (1.14) is satisfied, the weight function w must be such that

$$\int_0^\infty w(x)\,dx < 1.$$

(iv) Doubly stochastic Poisson processes

A different generalization of the Poisson process is obtained by supposing that there is a real-valued non-negative stochastic process $\{\Lambda(t)\}$, of pre-assigned structure and typically not observed, such that

$$\rho(t; \mathscr{H}_t, \mathscr{H}_t^\Lambda) = \lim_{\delta \to 0+} \delta^{-1} \mathrm{pr}\{N(t, t+\delta) > 0 | \mathscr{H}_t, \Lambda(s)$$
$$= \lambda(s)(-\infty < s \leqslant t)\}$$
$$= \lambda(t). \qquad (1.15)$$

Here $\mathscr{H}_t^\Lambda$ denotes the history of the process Λ at time t; thus (1.15), together with (1.11), specifies that the point process is a time-dependent Poisson process with a rate that is itself changing in time in accordance with a probabilistic law.

For example, the point process may be that of breakdowns of a loom. These occur irregularly in a way that is close to a Poisson process over fairly short time periods; however, the physical properties, such as relative humidity and quality of raw material, which affect the rate of occurrence, vary stochastically.

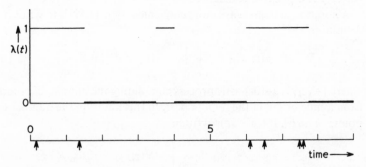

Fig. 1.4. *Special form of doubly stochastic Poisson process.* $\lambda(t)$, *unobserved rate process,* $\uparrow$, *points of point process.*

The full definition of the doubly stochastic Poisson process consists of (1.15) and (1.11) plus a specification of the properties of the stochastic process $\{\Lambda(t)\}$. It might often be a stationary process. A very special case is illustrated in Fig. 1.4, where the process $\{\Lambda(t)\}$ alternates stochastically between the levels 0 and 1. That is, the final point process, which is all that would typically be observed, consists of sections of a Poisson process of unit rate alternating with sections which are without points. A possible physical interpretation is that a component of some system has periods of use and of idleness and that failure can occur only during a period of use.

An alternative name for a doubly stochastic process is a process with random environment.

Now in principle this process can be expressed in terms of $\mathcal{H}_t$ only, in fact

$$\rho(t; \mathcal{H}_t) = E\{\Lambda(t)|\mathcal{H}_t\} \tag{1.16}$$

specifies the complete intensity function (1.10). Unfortunately, however, even in quite simple special cases the evaluation of the conditional expectation in (1.16) is difficult.

1.3 The specification and properties of point processes

The special processes introduced in Section 1.2 have illustrated a number of the general ideas required for the study of point processes. For the Poisson process, we have seen that the following specifications are equivalent:

(i) via the complete intensity function $\rho(t; \mathcal{H}_t)$;

(ii) via the joint distribution of the intervals between successive points, the interval specification;

(iii) via the joint distribution of the numbers of points in arbitrary sets $A_1, A_2, \ldots$, the counting specification.

We shall consider for most of the book, and throughout the present section, processes for which (1.11) holds, i.e. for which

$$\mathrm{pr}\{N(t, t + \delta) > 1\} = o(\delta).$$

For such processes each of the specifications (i)–(iii) is enough to define the process.

It is sometimes useful in constructing models for particular situations to introduce dependencies on observed explanatory variables denoted at time t by the vector $z(t)$. Suppose that in addition there is an unobserved stochastic process $\{\Lambda(t)\}$ of pre-assigned structure influencing the occurrence of points. Then an extremely general formulation is in terms of

$$\rho(t; \mathscr{H}_t, \mathscr{H}_t^z, \mathscr{H}_t^\Lambda) = \lim_{\delta \to 0+} \delta^{-1} \text{pr}\{N(t, t+\delta) > 0 \,|\, \mathscr{H}_t, \mathscr{H}_t^z, \mathscr{H}_t^\Lambda\},$$

(1.17)

where, for example, $\mathscr{H}_t^z$ gives $z(s)$ $(-\infty < s \le t)$. Note the special character of the one-sided dependence on time. We have already had as very special cases of (1.17), Poisson processes in which the rate is proportional to a scalar $z(t)$ and the doubly stochastic Poisson process in which the rate is equal to $\Lambda(t)$. The simplest processes, however, are defined in terms only of the history $\mathscr{H}_t$ of the point process itself and so in a general sense are self-exciting. Most, although not all, of the discussion will concern these.

From the point of view of general theory, it is sometimes convenient to regard the counting specification as central and to consider a point process to be defined by the joint distributions of counts in arbitrary sets; these joint distributions may depend on any explanatory variables available. This definition is in line with the general definition of a stochastic process as being determined by the joint distributions of the process at all possible sets of time instants, subject to selfconsistency. In applications to special processes this is rarely a convenient starting point, in connection with either stochastic processes generally or point processes in particular. This is clear from the discussion of Section 1.2 and in particular from the study of the Poisson process. Here both the intensity specification and the interval specification give insight into the structure of the process and into direct ways of constructing realizations.

Many of the problems to be studied amount to interrelating the different specifications. For instance, if a process is conveniently specified by an intensity, what can be said about the intervals between successive points and about the counts?

Very often the Poisson process serves as a basis for qualitative comparison. Because relative scatter is a first descriptive aspect of the form of a distribution, it is useful to define for the interval X between successive points, and for the count $N(t)$ of the number of points in $(0, t]$, indices of dispersion

$$I_X = \frac{\text{var}(X)}{\{E(X)\}^2}, \quad I(t) = \frac{\text{var}\{N(t)\}}{E\{N(t)\}}.$$

(1.18)

Because of well known properties of the exponential and Poisson distributions, these indices are both unity for the Poisson process. We use the terms overdispersion and underdispersion in an obvious way.

One of the first things to examine for a particular process is very often the nature of the indices (1.18).

Two important random variables associated with a point process are the forward and backward recurrence times measured from t. These are respectively the time measured forward from t to the next point, and the time measured backwards from t to the immediately preceding point, the latter being defined as zero if there is a point at t. The forward recurrence time arose naturally in Section 1.2 in connection with the definition of a renewal process.

Sometimes it is useful to examine the probability of a point in $(t, t + \delta)$, given that a point occurs at the origin. This is specified by the conditional intensity function

$$h(t) = \lim_{\delta_1, \delta_2 \to 0+} \delta_2^{-1} \mathrm{pr}\{N(t, t + \delta_2) > 0 | N(-\delta_1, 0) > 0\}. \quad (1.19)$$

For the renewal process, this gives the probability that in an ordinary renewal process a point occurs in $(t, t + \delta)$; note especially that the point in $(t, t + \delta)$ is not necessarily the first point after the origin.

One interesting family of problems, to be studied in Chapter 4, concerns the effect on the properties of one or more point processes of various transformations applied to the points.

Most but not all of the processes we consider are stationary in the sense that the joint distribution of $N(A_1)$, $N(A_2)$, ..., $N(A_k)$ is the same as that of $N(A_1 + \tau)$, $N(A_2 + \tau)$, ..., $N(A_k + \tau)$ for all τ and $k = 1, 2, \ldots$, where $A + \tau$ is the set of time values formed by translating A by τ. As the discussion of renewal processes shows, care is needed in studying the interval properties of stationary processes by asynchronous sampling, or in translating a specification in terms of counts into one based on intervals; the initial conditions, i.e. the distributions of the intervals immediately following the time origin, have to be defined appropriately.

1.4 Some generalizations

The situation discussed so far of points occurring singly in time can be generalized in many ways. There follow notes on some of these generalizations, together with brief comments on the nature of the new ideas involved.

(i) *Multiple occurrences*

The special examples introduced in Section 1.2 are all such that points occur in isolation, expressed formally by (1.11). In applications, however, we may well have multiple occurrences. The general theory can be set out to cover this possibility. In specific applications, it will usually be simplest to redefine a point to correspond to one or more occurrences at a time instant, and then to define the multiplicity of the point as the number of occurrences there.

A particular rather simple case arises when the multiplicities M of the different points are independent and identically distributed random variables, independent of the process of points. Then, if in a set A there are $N(A)$ points, with multiplicities $M_1, \ldots, M_{N(A)}$, the total number of occurrences in A is

$$R(A) = M_1 + \ldots + M_{N(A)},$$

with probability generating function

$$\begin{aligned}
G_{R(A)}(z) &= E(z^{R(A)}) \\
&= E_{N(A)} E\{z^{R(A)} | N(A)\} \\
&= G_{N(A)}\{G_M(z)\},
\end{aligned} \tag{1.20}$$

where G_M is the probability generating function for the multiplicity of a single point and $G_{N(A)}$ refers to the number of points in A. In particular, if the points follow a Poisson process of rate ρ and if $|A| = t$, then $N(A)$ has a Poisson distribution of mean ρt and

$$G_{R(A)} = \exp\{-\rho t + \rho t G_M(z)\}, \tag{1.21}$$

which corresponds to a compound Poisson distribution; see Feller (1968, Section 12.2).

Of course, many richer possibilities arise if the multiplicities can depend on features of the point process.

(ii) *Multivariate processes*
In most of the examples of Section 1.2 the points are indistinguishable except by their position in time. An important generalization arises if there are points of several different classes or types; we then call the process multivariate. The number of types may be finite or countably infinite. For the moment we consider there to be just two classes of point, and denote the numbers of points of the two classes in a time set A by $\{N^{(1)}(A), N^{(2)}(A)\}$. The full counting specification requires not only the joint distribution of $\{N^{(1)}(A_1), N^{(2)}(A_2)\}$ for arbitrary sets A_1 and A_2 but the joint distribution of all collections of such pairs.

If the histories at time t of the points of the two classes are $\mathcal{H}_t^{(1)}$ and $\mathcal{H}_t^{(2)}$, the complete intensity functions for the process are

$$\rho^{(1)}(t; \mathcal{H}_t^{(1)}, \mathcal{H}_t^{(2)}) =$$
$$\lim_{\delta \to 0+} \delta^{-1} \mathrm{pr}\{N^{(1)}(t, t+\delta) > 0, N^{(2)}(t, t+\delta) = 0 | \mathcal{H}_t^{(1)}, \mathcal{H}_t^{(2)}\}, \quad (1.22)$$

$$\rho^{(2)}(t; \mathcal{H}_t^{(1)}, \mathcal{H}_t^{(2)}) =$$
$$\lim_{\delta \to 0+} \delta^{-1} \mathrm{pr}\{N^{(1)}(t, t+\delta) = 0, N^{(2)}(t, t+\delta) > 0 | \mathcal{H}_t^{(1)}, \mathcal{H}_t^{(2)}\}, \quad (1.23)$$

$$\rho^{(12)}(t; \mathcal{H}_t^{(1)}, \mathcal{H}_t^{(2)}) =$$
$$\lim_{\delta \to 0+} \delta^{-1} \mathrm{pr}\{N^{(1)}(t, t+\delta) > 0, N^{(2)}(t+\delta) > 0 | \mathcal{H}_t^{(1)}, \mathcal{H}_t^{(2)}\}. \quad (1.24)$$

Here, while we continue to assume that, for the two classes of point separately, the condition (1.11) excluding multiple occurrences still holds, we do allow in accordance with (1.24) the possibility of a class 1 and a class 2 point occurring together. Alternatively, of course, we could regard such a double occurrence as a point of a new class, say class 3.

The properties of multivariate point processes are studied systematically in Chapter 5. Specification of suitable complete intensity functions allows the possibility of a rich variety of dependencies between the points of different classes or types.

(iii) Marked processes
Both generalizations (i) and (ii) are special cases of the notion of a marked point process, i.e. a point process in which a real-valued random variable, or vector of random variables, called a mark, is attached to each point. In (i) the mark is the multiplicity whereas in (ii) it is an indicator of the class of point. In the neurophysiological example mentioned in Section 1.1 the mark could be the magnitude of the peak signal at the point in question.

(iv) Spatial and multidimensional processes
In most of the discussion so far the points occur in time. If the points are arranged along a line in space, or more generally along a curve, all the same formal ideas apply. However, unless there is a clear physical reason for distinguishing the direction left to right as more natural for the process than the direction right to left, the complete intensity function in the form (1.10), while mathematically defined, is not really physically appropriate. A more symmetrical analogue of the complete intensity function is

$$\rho(t; \mathcal{E}_t) = \lim_{\delta_1, \delta_2 \to 0+} (\delta_1 + \delta_2)^{-1} \mathrm{pr}\{N(t-\delta_1, t+\delta_2) > 0 | \mathcal{E}_{(t-\delta_1, t+\delta_2)}\},$$
$$(1.25)$$

where $\mathscr{E}_{(a,\,b)}$ denotes all points occurring outside $(a,\,b]$, and $\mathscr{E}_t$ denotes all points excluding only any point occurring at t.

Similarly for points distributed in some higher dimensional space, such as the Euclidean plane or Euclidean space of three dimensions, we denote by $\mathscr{E}_{\delta_t}$ the positions of all points excluding those in a region δ_t centred on t and of measure (i.e. area, volume, etc.) $|\delta_t|$; there will usually be no real loss of generality in taking the small region to be an open ball. Then we define

$$\rho(t;\mathscr{E}_t) = \lim_{\delta_t \to 0} |\delta_t|^{-1} \mathrm{pr}\{N(\delta_t) > 0 | \mathscr{E}_{\delta_t}\}, \qquad (1.26)$$

where $N(\delta_t)$ is the number of points in δ_t and $\delta_t \to 0$ means that the diameter of the set δ_t, i.e. the greatest distance between two elements of δ_t, tends to zero. The Poisson process is defined by the requirement that (1.26) is constant. It is then easily shown that the counting properties of the Poisson process hold as in one dimension.

In one dimension a point process is often very conveniently specified theoretically or numerically via the intervals between successive points. In two or more dimensions, for an arbitrary point we can define its nearest neighbour, second nearest neighbour and so on, and specify these, for example using polar co-ordinates. It is clear, however, that the very simple specification of the whole process via successive intervals is lost in more than one dimension and, therefore, that the interval specification is much less important than in one-dimensional space.

It will be clear from the context what space is being considered, but in general the unqualified term point process will mean a process in one dimension in which the one-sided intensity function (1.10) is useful; the great majority of the book concerns this case. If we have a process specifically in 'space' rather than in time and for which the natural intensity function is (1.25) or (1.26) we call the process spatial. If points are positioned in 'space' and 'time', then we call the process spatial-temporal. Indeed most 'spatial' processes are reasonably thought of as generated in time, even if this is not explicitly recorded.

(v) Non-stationary and finite processes

The majority of the processes considered in this book are defined over an infinite time range and most are stationary in a sense to be explained in Section 2.2. Non-stationary processes can arise in particular when the rate of occurrence changes systematically in time; an important special case is when the total number of points is finite. We can illustrate such processes briefly in two ways, both connected with the Poisson process.

First, in the time-dependent Poisson process of rate $\rho(u)$, the number of points in $(0, t]$ can be shown to have a Poisson distribution of mean

$$\int_0^t \rho(u)\,du,$$

and therefore the total number of points in $(0, \infty)$ is finite with probability one if and only if

$$\int_0^\infty \rho(u)\,du < \infty.$$

In this case, if W denotes the time of occurrence of the last point, then

$$\mathrm{pr}(W < w) = \mathrm{pr}\{N(w, \infty) = 0\} = \exp\left\{-\int_w^\infty \rho(u)\,du\right\}, \quad (1.27)$$

so that the probability density function of W is, on differentiating (1.27),

$$f_W(w) = \rho(w)\exp\left\{-\int_w^\infty \rho(u)\,du\right\}. \quad (1.28)$$

Secondly, consider an ordinary Poisson process of rate ρ and suppose that after a random time L, the process of points is terminated. The random variable L is typically not observed, so that the point process is defined for all $t > 0$. The simplest situation arises when L has a distribution which is independent of the points in the originating Poisson process.

The total number, $N(0, \infty)$, of points in $(0, \infty)$, given that $L = l$, has a Poisson distribution of mean ρl. Thus the unconditional probability generating function of this number is

$$\int_0^\infty \exp\{\rho l(z - 1)\}f_L(l)\,dl = f_L^*(\rho - \rho z), \quad (1.29)$$

where $f_L^*(s) = E(e^{-sL})$ is the moment generating function of L. It follows in particular that if L has an exponential distribution the distribution of $N(0, \infty)$ is geometric. Further, unless L is constant, the distribution of $N(0, \infty)$ is overdispersed relative to the Poisson distribution.

(vi) *Discrete time processes*

A final generalization is to processes occurring in discrete time. These will not be treated in detail in the present book. No essentially new ideas are involved in passing to discrete time. Analytical solution tends to be simpler in continuous time, although for some problems in which analytical solution is cumbersome, discretization of the time scale offers one approach for computation.

Bibliographic notes, 1

References required for further development of particular points in the discussion are given in the text. The object of these and subsequent Bibliographic notes is to give some general references and brief historical remarks.

One of the earliest studies of the exponential distribution in a point process was in connection with free path length in the kinetic theory of gases (Clausius, 1858). Erlang, in his pioneer work on congestion in telephone systems, made extensive use of the Poisson process; for a historical review, see Jensen (1948). Khintchine (1960), in an investigation of queueing problems, made a careful study of point processes, or streams, considered as input processes. Applications in physics were studied by Bhabha and Heitler (1937); see in addition Srinivasan (1969). For applications in neurophysiology, see Fienberg (1974) and Sampath and Srinivasan (1977), and for models of earthquake occurrences, see Vere–Jones (1970).

Renewal theory first arose in deterministic studies of so-called self-renewing aggregates. There has been very extensive study of general renewal theory with emphasis on the analytical problems involved (Smith, 1958; Feller, 1971). For a much more applied account, see Cox (1962).

Developments both of the theory and of the applications of point processes up to the early 1970s were reviewed systematically in the book of conference papers edited by Lewis (1972). Daley and Milne (1973) have given a valuable bibliography up to about the same time. Subsequent accounts are by Srinivasan (1974) and Brillinger (1978a), the latter emphasizing the formal connections with time series analysis. A bibliography on the clustering of points in time and space is given by Naus (1979). For references on the general theory, see Bibliographic notes, 2.

Further results and exercises, 1

1.1. Discuss how to simulate a Poisson process of rate ρ starting from a source of random numbers uniformly distributed on $(0, 1)$.

Extend the procedure to simulate a time-dependent Poisson process where the rate function is specified numerically. [Section 1.2(i).]

1.2.   Obtain and solve the forward equations, referred to before (1.7), for the distribution of the number of points in $(0, t]$ in a Poisson process. Study first the time-homogeneous Poisson process and then the time-dependent Poisson process. [Section 1.2(i).]

1.3.   Prove that in a Poisson process of rate ρ the time up to the rth point has the density $\rho(\rho x)^{r-1} e^{-\rho x}/(r-1)!$, by the following different approaches:

   (i) obtain the moment generating function of the sum of r exponential variables;
   (ii) evaluate the convolution integrals associated with (i);
   (iii) note that the argument in (1.8) could be reversed;
   (iv) divide $(0, x]$ into a large number of equal subintervals and consider the probability that the last one contains a point and that of the remaining subintervals exactly $r-1$ contain points. [Section 1.2(i).]

1.4.   The rate of a Poisson process is $\rho(t) = \rho (2m \le t < 2m+1)$; $\rho(t) = 0$ $(2m+1 \le t < 2m+2)$; $m = 0, 1, \ldots$.   A sampling interval of unit length is selected at random over a long interval. Prove that the probability that the sampling interval contains n points is

$$\rho^{-1} \int_0^\rho \frac{e^{-u} u^n}{n!} \, du$$

and that the expected value and variance of the number of points are respectively $\frac{1}{2}\rho$ and $\frac{1}{2}\rho + \frac{1}{12}\rho^2$. [Sections 1.2(i), (iv) and 3.3(iii).]

1.5.   Obtain the moments of the length-biased distribution $zg(z)/\mu$ in terms of those of the underlying common density g of the intervals in a point process. Examine the relationships between (i) the means of the two distributions and (ii) the coefficients of variation of the two distributions. Discuss how from a large number of observations, effectively determining the length-biased density, one could determine the density g. [Section 1.2(ii); Cox (1969).]

1.6.   Prove that the equilibrium density (1.9) of the forward recurrence time is the same as the originating common density g for the intervals in a point process, if and only if both are exponential. Does this imply that the process is a Poisson process? Obtain a general relation between the moment generating functions of the density (1.9) and of g. [Section 1.2(ii).]

1.7. Examine the form taken by the compound Poisson distribution (1.21) when M has

  (i) a geometric distribution;

  (ii) a two-point distribution, taking values one or two.

[Section 1.4(i); Johnson and Kotz (1969), Chapter 8.]

1.8. If the random variables (M_1, M_2, M_3) have independent Poisson distributions of means (v_1, v_2, v_3), then $N_1 = M_1 + M_3$, $N_2 = M_2 + M_3$ are said to have a bivariate Poisson distribution. Obtain the joint probability distribution of N_1 and N_2 and the joint probability generating function $E(z_1^{N_1} z_2^{N_2})$; prove that the joint cumulant generating function is

$$\log E(e^{-s_1 N_1 - s_2 N_2}) = v_1(e^{-s_1} - 1) + v_2(e^{-s_2} - 1) + v_3(e^{-s_1 - s_2} - 1).$$

[Sections 1.4(ii), 5.3(i), 6.5(iii).]

1.9. Points are distributed in a Poisson process of rate ρ in a connected region of area a_0 in the plane. What is the probability that the region contains no points? Start from an arbitrary origin and search the region in a systematic way, e.g. in concentric circles of ever increasing radius, except where constrained by the boundary. Let A_1 be the area searched before encountering the first point. Prove that A_1 has an exponential distribution with parameter ρ, with an obvious modification if the region contains no points. Assuming that $\rho a_0 \gg 1$, prove that if A_k is the area searched up to the kth point, then A_1, $A_2 - A_1, A_3 - A_2, \ldots$ are independently and identically distributed.

   Generalize the results to more than two dimensions. [Sections 1.4(iv), 6.2(i).]

1.10. Prove from (1.29) that determination of the distribution of $N(0, \infty)$ is not enough to fix ρ and the distribution of L, but that if ρ and the distribution of $N(0, \infty)$ are both known, that of L can be determined. [Section 1.4(v).]

1.11. Prove the equivalence of the classes of finite processes formed from Poisson processes by termination

   (i) after a time exponentially distributed, independently of the original process; and

   (ii) after a number of points geometrically distributed, independently of the original process.

Find for such a process, with known parameters, the conditional probability that there are no points after time t given the positions of all points in $(0, t]$. [Section 1.4(v); Vit (1974).]

CHAPTER 2

Theoretical framework

2.1 Some basic definitions

In Chapter 1 point processes were introduced as stochastic processes for which each realization consists of a collection of points, each point having a well-defined position, usually in one-dimensional space, but possibly in some higher dimensional space. Such a process can be specified mathematically in several ways, in particular via the joint distributions of the counts of points in arbitrary sets or, if the process has no multiple simultaneous occurrences and is in one-dimensional space, via the joint distributions of intervals between successive points starting from a suitable origin, or via a complete intensity function. In applied work we use whichever of these approaches is convenient and feel free to pass between what are clearly physically equivalent specifications. For a unified mathematical development, however, it is necessary to choose a set of starting assumptions and to develop the theory in a connected way from these. In such a development, it will, for instance, be necessary to prove the existence of functions whose existence on physical grounds is obvious and to prove carefully results which are physically clear. Thus if a counting specification is taken as primary, the existence of a complete intensity function has to be proved; yet its existence is genuinely obvious, in the sense that without it the process cannot be simulated recursively in time, and hence in a practical sense is not defined.

In the present book we concentrate on aspects that are likely to be directly useful in applications, and hence put little emphasis on existence proofs and the like.

Mathematically, the most convenient starting point for a discussion of point processes is to regard N as a random counting measure; that is for any set A on the real line $\mathbb{R}$, $N(A)$ is a random nonnegative integer representing the number of points in A, these numbers being subject to obvious consistency requirements, namely that if A is the union of disjoint sets $\tilde{A}_1, \tilde{A}_2 \ldots$, then $N(A) = \Sigma N(\tilde{A}_i)$.

The joint distributions

$$\text{pr}\{N(A_i) = n_i; \ i = 1, \ldots, k\} \qquad (2.1)$$

are then specified in a consistent fashion for $n_i = 0, 1, 2, \ldots$; $i = 1, \ldots, k; k = 1, 2, \ldots$, where the A_i are arbitrary sets in $\mathbb{R}$. In fact to avoid consideration of meaningless events, the A_i are restricted to be so-called Borel sets, here and throughout the book.

If 0 is an arbitrary origin chosen without regard to the realization of the point process, then, provided that the multiplicities of simultaneous occurrences are finite, the points may be labelled according to the convention $\ldots \leq t_{-1} \leq t_0 \leq 0 < t_1 \leq t_2 \leq \ldots$. An alternative specification of the point process is thus to consider a consistent set of joint probability distributions for the sequence $\{T_i; \ i = 0, \pm 1, \ldots,\}$, where T_i is the random variable corresponding to an observed point t_i. If the process has no multiple points, so that the t_i are distinct, then the differences $\{t_i - t_{i-1}; \ i = 0, \pm 1, \ldots\}$ are the intervals between points, although to specify the process completely we need to know the position of the sequence relative to the origin. Therefore the process may be defined in terms of a consistent set of joint probability distributions for the sequence $\{T_1, T_i - T_{i-1}; \ i = 0, \pm 1, \ldots\}$.

There is also the important possibility discussed in Chapter 1, and mentioned again briefly above, of specifying a process without multiple occurrences via the complete intensity function (1.10).

In theoretical work on point processes it is usual to require that the measure N is finite, i.e. that with probability one the sequence $\{T_i\}$ has no finite limit points. In one sense this is a practically realistic requirement in that it will be possible to record only a finite number of points in a finite time. Nevertheless occasionally it may be useful in applications to use an idealized model in which an 'explosion' can take place, generating an infinite number of points in a finite time. A simple, very special, example is the time-dependent Poisson process of Section 1.2(i) with the rate function

$$\rho(t) = (t - [t])/(1 - t + [t]),$$

where $[t]$ is the integer part of t. The number of points in the interval

Table 2.1. *Realization of Poisson process on* $(0, 1]$ *with rate* $t/(1 - t)$. *Positions of first 20 points to 6 decimal places.*

0.515 087	0.593 880	0.759 121	0.801 176	0.986 752
0.999 517	0.999 771	0.999 856	0.999 918	0.999 954
0.999 991	0.999 996	0.999 997	0.999 998	0.999 998
0.999 999	1.000 000	1.000 000	1.000 000	1.000 000

$(a, 1 - \varepsilon)$ for $0 < a < 1$, $0 < \varepsilon < 1 - a$ has a Poisson distribution of mean

$$\int_a^{1-\varepsilon} \rho(t)dt = \int_a^{1-\varepsilon} \frac{t}{1-t}dt$$

and this tends to ∞ as $\varepsilon \to 0+$. Thus the number of points in $(a, 1]$ is infinite with probability one. Table 2.1 gives some idea of a typical realization. Nevertheless, from now on, we shall without further comment assume that N is finite.

One advantage of taking the counting specification as primary is that it extends immediately to processes in several dimensions. For particular processes in one dimension there is often much interest in intervals between successive points. The connection between counts and intervals is made explicit by relations of the form

$$\mathrm{pr}\{N(t) > n\} = \mathrm{pr}(T_{n+1} \leq t), \tag{2.2}$$

where, as before, $N((u, v])$ is written as $N(u, v)$, and $N(0, t)$ is abbreviated to $N(t)$; this expresses the fact that there are more than n points in $(0, t]$ if and only if the $(n + 1)$st point after the origin occurs by time t. Equations similar to (2.2) are used extensively in developing properties of point processes.

The interval approach presupposes a process on the real line and there is no immediate generalization for processes in higher dimensional spaces. Although, starting from an arbitrary origin, we may measure the distances to the nearest, next nearest, ..., kth nearest point, ..., obtaining a sequence $\{r_k; k = 1, 2, \ldots\}$, clearly these distances are not sufficient to determine the process. Incidentally, this is true even on the real line $\mathbb{R}$ unless the distances are signed. If the interval between two points in $\mathbb{R}$ is small then the points are close together, whereas if $r_k - r_{k-1}$ is small, the kth and $(k - 1)$st nearest points may nevertheless be widely separated spatially. One important use of the nearest neighbour distances r_k in higher dimensions is in the statistical analysis of spatial point processes.

It is fairly clear that the full set of probabilities (2.1) will be determined by suitable much more special sets of probabilities. For example, if N is finite and if there are no multiple occurrences, it can be shown (Oakes, 1972; Kallenberg, 1973) to be enough to specify (2.1) for $k = 1$, $n_1 = 0$ and for sets A_1 that are finite unions of half-open intervals. Essentially this is because by taking a sufficiently fine network of intervals we can ensure that at most one point falls in any such interval and the generalized addition law of probability can be applied to examine more complicated configurations.

2.2 Stationarity

The qualitative idea of stationarity is essentially the same for point processes as for stochastic processes generally. That is, the structure of the process has to be unaffected by translation of the time axis. The most general definition is that all the probabilities (2.1) should be unchanged by translating the sets $A_1, A_2, \ldots, A_k$ by an arbitrary shift τ. When this is so, the process is called completely (or strictly) stationary. Many weaker forms of stationarity have been defined, for example by restricting the sets $A_1, \ldots, A_k$ to be intervals and the integer k to be not more than some finite integer m, for instance $m = 1, 2$. In general the full power of complete stationarity is seldom required and it is usually sufficient in investigating a particular property of a process to assume that particular property to be stationary. Often we shall use the term stationary without qualification and in these instances the assumption being made should be clear from the context. Two particular forms of stationarity are commonly given specific names. If the distribution of $N(I)$ is invariant under translations of the arbitrary interval I then the process is simply stationary, whereas if the mean and variance of $N(A)$ are invariant under translations of the arbitrary set A then the process is weakly stationary.

Having considered stationarity in terms of counts, we need also to look at the interval sequence. A point process is said to have stationary intervals if the joint distribution of

$$X_{s_1}, X_{s_2}, \ldots, X_{s_k} \tag{2.3}$$

depends only on $s_2 - s_1, \ldots, s_k - s_1$, for all $s_1, \ldots, s_k$ and all k, where $X_i = T_i - T_{i-1}$ is the interval between the $(i-1)$st and ith point. A simple example is a renewal process, defined as having independent and identically distributed intervals.

Description of the link between stationarity defined in terms of counts and the property of stationary intervals needs a little care. First consider a completely stationary process and imagine a new origin chosen as uniformly distributed over some long period, the length of that period tending to infinity. Now it is clear that there is a relatively high chance that the origin will fall in a long interval between successive points, in fact that length-biased sampling will operate; see Section 1.2(ii). Because intervals between successive points are in general dependent, it follows that the interval between the first and second points following the time origin will in general be atypical; except for special processes there will be an analogous distortion of subsequent intervals. That is, the sequence of intervals measured from an arbitrary time origin is typically not stationary; it does, however, have distributional properties in principle determined by the counting specification (2.1). Conversely if we start with the property of stationary intervals, we are working in effect with measurements defined from an arbitrary point as origin, and the resulting counting process will not in general be stationary.

To study the interval sequence the most useful approach is to consider a completely stationary process and then to examine properties given a point at the origin; mathematically this is achieved by conditioning on the event $N(-\delta, 0) > 0$ in the limit as $\delta \to 0+$. The interval sequence so defined is then stationary. This approach is discussed in greater detail in Section 2.4.

2.3 Orderliness

It is simplest to consider point processes consisting of a sequence of single points and in Chapter 1 the condition that, as $\delta \to 0+$,

$$\text{pr}\{N(t, t+\delta) > 1\} = o(\delta), \quad \text{all} \quad t \in \mathbb{R}, \tag{2.4}$$

has been imposed with no explanation other than the statement that (2.4) prevents the process having multiple simultaneous occurrences. A process satisfying (2.4) will be called orderly, whereas a process for which

$$\text{pr}\{N(\{t\}) > 1 \text{ for some } t \in \mathbb{R}\} = 0, \tag{2.5}$$

where the set $\{t\}$ consists of the singleton $t \in \mathbb{R}$, is said to have no multiple simultaneous occurrences. In fact (2.4) can be shown to imply (2.5) (Daley, 1974). Alternative definitions of orderliness are sometimes used, but in this book it will be sufficient to consider only the above two and, in fact, for most of the processes discussed the definitions are equivalent, so that the terms orderly and without multiple occurrences can be used interchangeably.

Suppose now that a process is simply stationary, but not necessarily orderly. Then the mean $E\{N(I)\}$, if finite, is proportional to the Lebesgue measure of I, i.e.

$$E\{N(I)\} = \rho |I|,$$

where ρ is called the intensity or rate of the process. Note that the rate ρ could be defined directly by

$$\rho = \lim_{\delta \to 0+} \delta^{-1} E\{N(\delta)\}. \tag{2.6}$$

Another constant of the process is the occurrence parameter v which is defined as the limit

$$v = \lim_{\delta \to 0+} \delta^{-1} \mathrm{pr}\{N(\delta) > 0\},$$

since $N(u, u + t)$ has the same distribution as $N(t)$, for any u, in a simply stationary process. It is intuitively clear that while ρ is the overall rate of the process including multiplicities, v is the rate of the process of instants at which points occur, ignoring their multiplicities. Therefore, $v \leq \rho$, with equality if the process has no multiple occurrences, a result known as Korolyuk's theorem (Khintchine, 1960; Leadbetter, 1968). In general

$$\rho = vE(M), \tag{2.7}$$

where M is the multiplicity of an arbitrary point.

A further feature is that if the process is simply stationary with finite rate ρ, then orderliness and the property of no multiple occurrences are equivalent, this result being known as Dobrushin's lemma (Leadbetter, 1968); also the equality $\rho = v$ implies the orderliness of the process (Zitek, 1957). Thus for a simply stationary process with finite ρ, (2.4) and (2.5) are equivalent and

$$\lim_{\delta \to 0+} \delta^{-1} \mathrm{pr}\{N(\delta) > 0\} = E\{N(0, 1)\}.$$

If, however, the process is not stationary the situation is more complicated (Leadbetter, 1972a).

As noted in Section 1.4(i), the simplest way in practice to deal with processes with multiple occurrences is to study first the process of instants of occurrences, regardless of multiplicity, and then to study the multiplicities.

Finally, we note that the condition that the limiting behaviour of $\delta^{-1} \mathrm{pr}\{N(t, t + \delta) > 0\}$ is independent of t, is weaker than simple stationarity.

2.4 Palm distributions

As noted in Section 2.2, the relation between complete stationarity of the counts and stationarity of the intervals between successive points is mathematically indirect, even though it is clear intuitively that the two ideas are virtually identical. The counting aspects of a process are described relative to an arbitrary time instant, the origin, while to investigate the sequence of intervals, $\{X_i\}$, between successive points in a process with no multiple occurrences, we need to consider the process starting from an arbitrary point of the process. To find the survivor function, $\mathscr{F}_X$, of a typical interval X of a stationary process from the distribution of counts, consider

$$\mathscr{F}_X(x) = \text{pr}(X > x) = \lim_{\delta \to 0+} \text{pr}\{N(0, x) = 0 | N(-\delta, 0) > 0\}, \quad (2.8)$$

which is the limiting probability that, given that a point occurs immediately before the origin, the next point of the process occurs after the instant x. Now, by stationarity,

$$\text{pr}\{N(0, x) = 0 \text{ and } N(-\delta, 0) > 0\}$$
$$= \text{pr}\{N(0, x) = 0\} - \text{pr}\{N(-\delta, x) = 0\}$$
$$= \text{pr}\{N(x) = 0\} - \text{pr}\{N(x + \delta) = 0\},$$

so that

$$\text{pr}\{N(0, x) = 0 | N(-\delta, 0) > 0\} \delta^{-1} \text{pr}\{N(\delta) > 0\}$$
$$= -\delta^{-1}[\text{pr}\{N(x + \delta) = 0\} - \text{pr}\{N(x) = 0\}]. \quad (2.9)$$

The limit

$$\lim_{\delta \to 0+} \delta^{-1} \text{pr}\{N(\delta) > 0\}$$

has already been defined as the occurrence parameter v of the process, which we assume is finite. If the distribution of $N(x)$ is given by

$$\text{pr}\{N(x) = k\} = p_k(x) \quad (k = 0, 1, \ldots), \quad (2.10)$$

then in the limit as $\delta \to 0+$, (2.9) becomes

$$\mathscr{F}_X(x)v = -D_x p_0(x), \quad (2.11)$$

where D_x denotes the derivative, or where necessary the right-hand derivative. Of course it follows from Section 2.3 that, since the process is stationary and has no multiple occurrences, we can replace v in (2.11) by the rate ρ of the process. Now $p_0(x)$ is the probability that, starting from an arbitrary time instant, there are no process points in the following interval of length x. This is equivalent to saying that the forward recurrence time from the arbitrary instant to the next point of the process is at least x. Thus equation (2.11) links the distribution of the interval between successive points, with survivor function $\mathscr{F}_X(x)$, to that of the forward recurrence time, with survivor function $p_0(x)$. We have already obtained equation (2.11) by an alternative argument and in a different notation, for the special case of a renewal process; see (1.9). There the common distribution of intervals in the renewal process had survivor function $\mathscr{G}(x)$ and mean μ_X, so that the rate of the process was μ_X^{-1}.

A simple but atypical case of these formulae is obtained for the Poisson process of rate ρ. Then $p_0(x) = e^{-\rho x}$, (2.11) leads to $\mathscr{F}_X(x) = e^{-\rho x}$, and the two survivor functions mentioned in the previous paragraph are identical. It follows immediately from (2.11) that $p_0(x) = \mathscr{F}_X(x)$ if and only if both are exponential. Of course this specifies only the marginal distribution of intervals between successive points and does not imply that the process is a Poisson process.

More general results connecting distributions of events 'conditional on a point at the origin' with those where the origin is an arbitrary instant may be obtained. We shall assume for the rest of this section that the point process to be considered is completely stationary, has a finite occurrence parameter v and is orderly, so that v is equal to the rate ρ of the process. Then, for each $x > 0$, the Palm distribution is a discrete distribution defined by

$$\pi_k(x) = \lim_{\delta \to 0+} \text{pr}\{N(0, x) = k \mid N(-\delta, 0) > 0\}, \qquad (2.12)$$

for $k = 0, 1, \dots$. In a careful mathematical development the existence of v and $\pi_k(x)$, and more generally of other limiting probabilities of events $\mathscr{B}$ given $N(-\delta, 0) > 0$, of the form

$$\lim_{\delta \to 0+} \text{pr}\{\mathscr{B} \mid N(-\delta, 0) > 0\} \qquad (2.13)$$

has to be proved. To prove rigorously that (2.13) defines a probability measure, a mathematically elegant approach is through the use of marked point processes. To each point of the process is attached a mark, namely the entire realization of the process 'centred' on that particular point as the origin. This device enables a probability measure Π to be defined for events $\mathscr{B}$ on those processes which have a point at the origin; essentially a 'random point' is being chosen as the origin. The measure Π can then be shown to satisfy

$$\Pi(\mathscr{B}) = \lim_{\delta \to 0+} \text{pr}\{\mathscr{B} \mid N(-\delta, 0) > 0\}$$

for a wide class of events $\mathscr{B}$. The measure Π is called the Palm measure of the process (Leadbetter, 1972a; Matthes, 1964).

In equation (2.11) the distribution of the interval measured from an arbitrary time instant to the next point of the process, is linked to that of the interval between successive points. In the same way the functions $\pi_k(x)$ defining the Palm distributions given in (2.12), which specify the distribution of the number of points in an interval of length x given a point at the origin, can be connected with the functions $p_k(x)$

which give the distribution of the number of points in an interval of length x which starts at an arbitrary origin. These connecting equations are known as the Palm–Khintchine equations, and may be derived as follows. Since the process is orderly, if $k > 0$, as $\delta \to 0+$

$$p_k(x + \delta) = \mathrm{pr}\{N(-\delta, x) = k\}$$
$$= \mathrm{pr}\{N(-\delta, 0) = 0, N(0, x) = k\}$$
$$\quad + \mathrm{pr}\{N(-\delta, 0) = 1, N(0, x) = k - 1\} + o(\delta)$$
$$= p_k(x) - \mathrm{pr}\{N(-\delta, 0) > 0, N(0, x) = k\}$$
$$\quad + \mathrm{pr}\{N(-\delta, 0) > 0, N(0, x) = k - 1\} + o(\delta),$$

so that

$$\delta^{-1}\{p_k(x + \delta) - p_k(x)\}$$
$$= -\delta^{-1}\mathrm{pr}\{N(-\delta, 0) > 0\}[\mathrm{pr}\{N(0, x) = k | N(-\delta, 0) > 0\}$$
$$\quad - \mathrm{pr}\{N(0, x) = k - 1 | N(-\delta, 0) > 0\}] + o(1)$$

and hence

$$D_x p_k(x) = -\rho\{\pi_k(x) - \pi_{k-1}(x)\}; \tag{2.14}$$

again D_x indicates the (possibly right-hand) derivative. The corresponding equation for $k = 0$ has already been derived in (2.11) and is

$$D_x p_0(x) = -\rho\pi_0(x). \tag{2.15}$$

The integral forms for (2.14) and (2.15) are

$$p_k(x) = -\rho \int_0^x \{\pi_k(u) - \pi_{k-1}(u)\} du \quad (k = 1, 2, \ldots), \tag{2.16}$$

$$p_0(x) = 1 - \rho \int_0^x \pi_0(u) du. \tag{2.17}$$

It follows from (2.14) and (2.15) that

$$-\frac{1}{\rho}D_x\{p_0(x) + \ldots + p_k(x)\} = -\frac{1}{\rho}D_x\mathrm{pr}\{N(x) \le k\} = \pi_k(x),$$

and therefore that the probability of having exactly k points in $(0, x]$, starting from an event at 0, can be obtained by differentiating the probability of getting no more than k points in $(0, x]$, when 0 is an arbitrary time instant. Alternatively, from (2.16) and (2.17),

$$p_0(x) + \ldots + p_k(x) = \mathrm{pr}\{N(x) \le k\} = 1 - \rho \int_0^x \pi_k(u) du, \tag{2.18}$$

so that the probability of getting not more than k points in $(0, x]$, when 0 is an arbitrary instant, can be obtained by integrating the probability of exactly k points in the interval when there is a point at 0. In addition, the right-hand side of (2.18) is equal to

$$\rho \int_x^\infty \pi_k(u)\,du = \rho \int_0^\infty \pi_k(y + x)\,dy,$$

so that we may justify (2.18) by the argument that if 0 is an arbitrary time instant and there are no more than k points in $(0, x]$, then there must exist a point with co-ordinate $-y$, for some $y > 0$, such that there are exactly k points in $(-y, x]$. Since the process is orderly, the probability of a point in $(-y, -y + \delta]$ is $\rho\delta + o(\delta)$ and therefore

$$\text{pr}\{N(x) \le k\} = \rho \int_0^\infty \pi_k(y + x)\,dy.$$

Equations (2.14) and (2.15) or (2.16) and (2.17) can be summarized using generating functions. For if we define

$$G(z; x) = \sum_{k=0}^\infty z^k p_k(x)$$

and

$$G_0(z; x) = \sum_{k=0}^\infty z^k \pi_k(x),$$

so that G refers to an arbitrary origin while G_0 refers to the situation given a point at the origin, then

$$D_x G(z; x) = -\rho(1 - z)G_0(z; x),$$

or equivalently

$$G(z; x) = 1 - \rho(1 - z)\int_0^x G_0(z; u)\,du.$$

A simple example is provided by a renewal process (see Section 1.2(ii)) in which the intervals between successive points are independently and identically distributed with density g. Then the Palm probabilities refer to a process starting with a point at, or just before, the origin, i.e. to an ordinary renewal process, whereas the probabilities $\{p_k(x)\}$ refer to a process starting from an arbitrary time origin, i.e. to an equilibrium renewal process. In this case the Palm probabilities are readily computed via (2.2): in fact

$$\sum_{l=k}^\infty \pi_l(x) = G^{(k)}(x),$$

where the right-hand side denotes the cumulative distribution of $X_1 + \ldots + X_k$, obtained by k-fold convolution. That is,

$$\pi_k(x) = G^{(k)}(x) - G^{(k+1)}(x)$$

and the $p_k(x)$ are given by (2.16) and (2.17). A different approach to these results, specific however to renewal processes, is via the Laplace transforms of the probability generating functions for the p_k and the π_k (Cox, 1962, p. 38).

2.5 Moments

In any study of random variables it is quite often the first and second order moments that are of most interest. Thus for a point process we may consider

$$E\{N(A)\}, \quad \text{var}\{N(A)\}, \quad \text{cov}\{N(A), N(B)\} \qquad (2.19)$$

for the counts in arbitrary sets A and B, and

$$E(X_i), \quad \text{var}(X_i), \quad \text{cov}(X_i, X_{i+j}), \qquad (2.20)$$

for the intervals between successive points.

For stationary orderly processes of finite rate ρ, it is clear that

$$E\{N(A)\} = \rho|A|, \quad E(X_i) = 1/\rho, \qquad (2.21)$$

where $|A|$ is the measure of the set A, so that the first order properties are essentially equivalent; the second result in (2.21) is easily proved rigorously under weak conditions by applying the strong law of large numbers to the sum of a large number of Xs.

The second order properties in (2.19) and (2.20) are, however, not simply related, except in an asymptotic sense to be developed later.

For disjoint sets A and B, it is clear that

$$2\,\text{cov}\{N(A), N(B)\} = \text{var}\{N(A \cup B)\} - \text{var}\{N(A)\} - \text{var}\{N(B)\}, \qquad (2.22)$$

so that the calculation of variances can be regarded as the crucial step. Incidentally, a decomposition analogous to (2.22) is not possible for higher moments; for example $E\{N(A)N(B)N(C)\}$ cannot be expressed in terms of the first three moments of single counts.

The simplest special case of (2.19) is obtained by taking A to be the interval $(0, t]$. For non-stationary processes we introduce the local rate $\rho(t)$, extending (2.6) and defined by

$$\rho(t) = \lim_{\delta \to 0+} \delta^{-1} E\{N(t, t + \delta)\},$$

and consider

$$P(t) = E\{N(t)\}, \quad V(t) = \text{var}\{N(t)\},$$

called respectively the mean- and variance-time functions. The index of dispersion $I(t) = V(t)/P(t)$ was introduced in (1.18) as providing some comparison with the Poisson distribution, for which $I(t) = 1$.

In general it follows from the additive property of expectations that

$$P(t) = \int_0^t \rho(u) du$$

and that, in particular, if the process is stationary with rate ρ, $P(t) = \rho t$.

For a stationary process, the function $V(t)$ is mathematically equivalent to, and usually most easily derived from, the conditional intensity function, h, of (1.19):

$$h(t) = \lim_{\delta_1, \delta_2 \to 0+} \delta_2^{-1} \text{pr}\{N(t, t + \delta_2) > 0 | N(-\delta_1, 0) > 0\};$$

thus h is closely associated with the Palm measure of Section 2.4. It is crucial to appreciate the distinction between h and the density of intervals between successive points. The latter is a probability density of a random variable, has total integral one, and specifies the first point after the one at the origin. The former function is not a probability density of a random variable and concerns the conditional probability that there is a point close to t regardless of whether it is the first, second, ... point. Thus for a stationary orderly process of rate ρ, we shall have $h(t) \to \rho$ as $t \to \infty$, unless the process has some very long-term memory.

To calculate the function $V(t)$ it is useful to write formally

$$N(t) = \int_0^t dN(z), \tag{2.23}$$

considered as a limit of sums of counts in small intervals. If we apply to (2.23) the general formula for the variance of a sum of random variables we can write

$$\text{var}\{N(t)\} = \int_0^t \text{var}\{dN(z)\} + 2 \iint_{\substack{0 < z < t \\ 0 < u \leq t - z}} \text{cov}\{dN(z), dN(z + u)\}, \tag{2.24}$$

where again the integral is to be defined as the limit of a sum. Note particularly that in the last term of (2.24) the variable u is strictly positive since the contribution to the total variance when u is zero has been separated into the preceding term. Study of integrals like (2.23) and (2.24) often arises in applications and the following account sets out carefully the arguments involved.

If the process is orderly then the probability of at least two points in $(z, z + \delta]$ is $o(\delta)$, so that in the limit as $\delta \to 0 +$ we may consider $N(z, z + \delta)$ as a variable taking only the values zero and one. Thus, for a stationary process,

$$
\begin{aligned}
\mathrm{var}\,\{N(z, z + \delta)\} &= E[\,\{N(z, z + \delta)\}^2\,] - [E\{N(z, z + \delta)\}\,]^2 \\
&= \mathrm{pr}\,\{N(z, z + \delta) = 1\} \\
&\quad - [\mathrm{pr}\,\{N(z, z + \delta) = 1\}\,]^2 + o(\delta) \\
&= \rho\delta + o(\delta),
\end{aligned}
\tag{2.25}
$$

while, for $u > 0$,

$$
\begin{aligned}
\mathrm{cov}\,&\{N(z, z + \delta_1),\, N(z + u, z + u + \delta_2)\} \\
&= E[N(z, z + \delta_1)E\{N(z + u, z + u + \delta_2)|N(z, z + \delta_1)\}\,] \\
&\quad - E\{N(z, z + \delta_1)\}E\{N(z + u, z + u + \delta_2)\} \\
&= \mathrm{pr}\,\{N(z, z + \delta_1) = 1\}\,\mathrm{pr}\,\{N(z + u, z + u + \delta_2) = 1|N(z, z + \delta_1) = 1\} \\
&\quad - \mathrm{pr}\,\{N(z, z + \delta_1) = 1\}\,\mathrm{pr}\,\{N(z + u, z + u + \delta_2) = 1\} + o(\delta_1\delta_2) \\
&= \rho h(u)\delta_1\delta_2 - \rho^2\delta_1\delta_2 + o(\delta_1\delta_2).
\end{aligned}
\tag{2.26}
$$

Therefore, using (2.25) and (2.26) in the limit as δ_1 and δ_2 tend simultaneously to zero, we have from (2.24) that

$$
\mathrm{var}\,\{N(t)\} = \int_0^t \rho\,dz + 2\int_0^t dz \int_0^{t-z} du\,\{\rho h(u) - \rho^2\}
$$

$$
= \rho t + 2\rho \int_0^t (t - u)h(u)du - \rho^2 t^2.
\tag{2.27}
$$

It follows from (2.27) that $V(t)$ can be written in the form

$$
\mathrm{var}\,\{N(t)\} = V(t) = \int_0^t dz \int_0^t du\, c(u - z),
$$

where, for $u \geq 0$,

$$
c(u) = \rho\delta(u) + \rho h(u) - \rho^2
\tag{2.28}
$$

with, because by stationarity $h(-u) = h(u)$,

$$
c(-u) = c(u),
$$

and where δ is the Dirac delta function. The delta function appears in

(2.28) for the same reason that the right-hand side of (2.24) is split into two parts and represents the contribution to the total variance $V(t)$ from the variance of each $N(\{u\})$ for $u \in (0, t]$. The function c is called the covariance density and knowledge of the conditional intensity h is clearly equivalent to knowledge of the function c. In particular, as $u \to \infty$ the conditions $h(u) \to \rho$ and $c(u) \to 0$ are equivalent. Intuitively, the interpretation of c is that $c(u)\delta_1\delta_2$ is approximately the covariance between the counts in two intervals, of lengths δ_1 and δ_2, a distance u apart, when δ_1 and δ_2 are small.

For the covariance of the counts in arbitrary sets A and B we may use directly an argument similar to that used above for $V(t)$, or we may work via (2.22), to obtain

$$
\text{cov}\{N(A), N(B)\} = \int_A dz \int_B du\, c(u - z)
$$

$$
= \rho|A \cap B| + \rho \int_A dz \int_B du\, h(u - z) - \rho^2|A||B|,
$$

where $A \cap B$ is the common part of the sets A and B.

This method generalizes immediately to more complex situations to be studied more thoroughly in Section 5.6. For example, suppose that we define a stochastic process $Y(t)$ which is generated by the point process N in the sense that each point of the process occurring before t adds a contribution to $Y(t)$ depending on how long before t the point occurred. Specifically define

$$
Y(t) = \int_{-\infty}^{t} w(t - z)dN(z) = \sum_{\{i\,:\,T_i \leq t\}} w(t - T_i), \qquad (2.29)
$$

where $\{T_i\}$ are the times of points of the process, and the function w is such that all the relevant integrals converge. Clearly, $Y(t)$ is related to the self-exciting process of Section 1.2(iii). Then

$$
E\{Y(t)\} = \rho \int_0^{\infty} w(u)du,
$$

and

$$
\text{var}\{Y(t)\} = \text{var}\left\{ \int_{-\infty}^{t} w(t - z)dN(z) \right\}
$$

$$
= \int_{-\infty}^{t} \int_{-\infty}^{t} w(t - z)w(t - u)\text{cov}\{dN(z), dN(u)\}
$$

$$= \rho \int_0^\infty w^2(u)\,du + 2\rho \int_0^\infty du \int_u^\infty dv w(u)w(v)h(v-u)$$

$$- \rho^2 \left\{ \int_0^\infty w(u)\,du \right\}^2 .$$

The autocovariance function of Y may be obtained similarly.

The second-order properties of a stationary sequence of intervals $\{X_i\}$ are summarized by the sequence of autocovariances

$$c_k = \mathrm{cov}(X_i, X_{i+k}) \quad (k = 0, 1, 2, \ldots). \tag{2.30}$$

Of course, because processes of intervals are typically far from Gaussian processes, the autocovariance sequence is at best a description of some aspects of any dependency that may be present. As already noted, the second-order properties of counts and of intervals are complementary; roughly speaking, the former are most likely to be revealing when any underlying structure progresses in time regardless of the occurrence of intervening points, whereas some processes, especially renewal processes as an extreme, are most aptly described in terms of intervals between successive points.

One rather artificial example of the usefulness and limitations of (2.30) is provided by a process in which successive intervals X_i and X_{i+1} between adjacent points are identical with probability p and independent with probability $1 - p$. Successive events of identity and independence are themselves mutually independent and independent of the magnitude of the realized interval. Thus a particular realized interval, x, occurs exactly r times with probability $p^{r-1}(1-p)$ $(r = 1, 2, \ldots)$. It is easily shown that $c_k = \mathrm{var}(X)p^k$, a particular illustration of the general result that Markov processes with essentially linear structure have geometrically decaying autocovariance (Cox and Miller, 1965, p. 289). The nature of the dependence is seen directly from the special form of the joint distribution of two or more intervals, with its concentration of probability along the unit line or plane.

If we consider the sum of intervals, $S_k = X_1 + \ldots + X_k$, then an index of dispersion I_k may be defined by

$$I_k = \frac{\mathrm{var}(S_k)}{k\{E(X)\}^2}, \tag{2.31}$$

where the standardization is chosen so that $I_k = 1$ for all k for a Poisson process. There is an asymptotic link between I_k and the index of dispersion $I(t)$ for the counting process defined in (1.18), namely that

$$\lim_{k \to \infty} I_k = \lim_{t \to \infty} I(t). \tag{2.32}$$

To prove this result, we assume that S_k is asymptotically normally distributed with a variance which is of order k, say $\text{var}(S_k) = ka_k$, where a_k is $O(1)$ as $k \to \infty$ and is bounded away from zero. It is convenient to assume that the process has a point at the origin, and then to use the relation

$$\text{pr}\{N(t) < k\} = \text{pr}(S_k > t)$$

with k and t connected by the equation

$$k = t/E(X) + y(ta_k)^{1/2}/\{E(X)\}^{3/2}$$

for fixed y. Then

$$\text{pr}\{N(t) < k\} = \text{pr}\left\{ \frac{S_k - kE(X)}{\sqrt{(ka_k)}} > -y\left(1 + y\sqrt{\left\{\frac{a_k}{tE(X)}\right\}}\right)^{-1/2} \right\},$$
$$\tag{2.33}$$

so that in the limit as t and k tend to infinity the right-hand side of (2.33) tends to $\Phi(y)$, where Φ is the cumulative normal distribution function. Note that, in fact, since k is an integer and y is fixed, if $k \to \infty$ then $t \to \infty$ along a subsequence of the reals. Thus $N(t)$ also is asymptotically normally distributed, with asymptotic variance to mean ratio

$$\lim_{t \to \infty} I(t) = \lim_{k \to \infty} \frac{a_k}{\{E(X)\}^2} = \lim_{k \to \infty} I_k.$$

Although we have concentrated on the first and second order properties both of counts and of intervals it is of course straightforward to define higher moments in both cases. For intervals, product moments of the form

$$E\{X_{j_0} X_{j_0 + j_1} \cdots X_{j_0 + \ldots + j_k}\} \quad (0 \le j_i; i = 1, \ldots, k; k = 0, 1, \ldots)$$

may be of interest. In the case of counts one may consider factorial moments of the form

$$E[N(A)\{N(A) - 1\}\{N(A) - 2\} \ldots \{N(A) - r\}] \quad (r = 0, 1, \ldots)$$

or, more generally, product moments of the form

$$E\{N(A_1) \ldots N(A_k)\} \quad (k = 1, 2, \ldots).$$

For processes without multiple occurrences it is sometimes convenient to define the product densities $p_k(t_1, \ldots, t_k)$ by

$$p_k(t_1, \ldots, t_k)dt_1 \ldots dt_k = \text{pr}\{dN(t_1) = \ldots = dN(t_k) = 1\},$$

where $t_1, \ldots, t_k$ are distinct and $k = 1, 2, \ldots$. These densities lead immediately to the factorial moments, since

$$\int dt_1 \ldots \int dt_k p_k(t_1, \ldots, t_k)$$
$$= E[N(A)\{N(A) - 1\} \ldots \{N(A) - k + 1\}],$$

where the integral is over distinct t_i, $t_i \in A$ $(i = 1, \ldots, k)$.
With more effort, the result extends to the general product moments.
For example, for arbitrary sets A, B

$$E\{N(A)N(B)\} = \int dt_1 \int_{\{t_1 \in A, t_2 \in B, t_1 \neq t_2\}} dt_2 p_2(t_1, t_2) + \int_{A \cap B} dt p_1(t).$$

Distinct processes may have the same set of product densities, essentially because moments may not uniquely determine a distribution.

2.6   Spectral properties

In the theory of stationary time series it is valuable to consider frequency domain analysis in parallel with time domain analysis. In second-order theory this means considering the power spectrum in parallel with the autocovariance function. We can therefore introduce spectra corresponding to the second-order analyses both of intervals and of counts. For the first we write

$$c_k = \int_{-\pi}^{\pi} e^{ik\omega} \phi(\omega) d\omega \tag{2.34}$$

for processes with a power spectral density ϕ, with the more general form

$$c_k = \int_{-\pi}^{\pi} e^{ik\omega} d\Phi(\omega)$$

available when there is a discrete spectrum. It follows from (2.34) that

$$\phi(\omega) = \frac{1}{2\pi} \sum_{k=-\infty}^{\infty} c_k e^{-ik\omega} = \frac{1}{2\pi}\left\{c_0 + 2\sum_{k=1}^{\infty} c_k \cos(k\omega)\right\} \tag{2.35}$$

because $c_k = c_{-k}$.

There is a parallel spectral decomposition of the process $\{X_j\}$ in the form

$$X_j = E(X) + \int_{-\pi}^{\pi} e^{ij\omega} dZ_X(\omega), \qquad (2.36)$$

where $\{Z_X(\omega)\}$ is a process of orthogonal increments such that

$$E\{dZ_X(\omega)d\bar{Z}_X(\omega')\} = \phi(\omega)\delta(\omega - \omega')d\omega d\omega',$$

and $\bar{Z}$ is the complex conjugate of Z.

This representation is, however, of limited usefulness because, in order to represent non-negative random variables $\{X_j\}$, severe restrictions on the full probability distribution of $Z_j(\omega)(-\pi \le \omega < \pi)$ are required, i.e. (2.36) is essentially a second-order decomposition only.

The parallel to (2.35) for the covariance density of counts, c, is obtained by writing

$$\psi(\omega) = \frac{1}{2\pi}\int_{-\infty}^{\infty} c(u)e^{-i\omega u}du \qquad (2.37)$$

$$= \frac{\rho}{2\pi} + \frac{\rho}{2\pi}\int_{-\infty}^{\infty} \{h(u) - \rho\}e^{-i\omega u}du.$$

For a Poisson process, and more generally for any process with $h(u) = \rho$, we have the 'white' spectrum $\psi(\omega) = \rho/(2\pi)$.

The Fourier inverse of (2.37) is

$$c(u) = \int_{-\infty}^{\infty} e^{iu\omega}\psi(\omega)d\omega, \qquad (2.38)$$

and again there is a second-order decomposition of the process given by

$$N(u, u + v) = \rho v + \int_{-\infty}^{\infty} \frac{e^{i(u+v)\omega} - e^{iu\omega}}{i\omega}dZ_N(\omega), \qquad (2.39)$$

where $\{Z_N(\omega)\}$ is a process of orthogonal increments such that

$$E\{dZ_N(\omega)d\bar{Z}_N(\omega')\} = \psi(\omega)\delta(\omega - \omega')d\omega d\omega'. \qquad (2.40)$$

Some use is made of the spectra in Section 4.4 in connection with the random translation of processes.

2.7 The probability generating functional

In the study of a non-negative integer-valued random variable Y, the probability generating function

$$G_Y(z) = E(z^Y) = E(e^{Y \log z})$$

is a particularly useful tool. The argument z can be taken so that $-1 < z < +1$. The function $G_Y(z)$ provides a complete description of the distribution of the random variable Y, because any property of Y can be deduced from $G_Y(z)$ and conversely. In particular the moments of Y are simply obtained from the derivatives of $G_Y(z)$ in the limit as $z \to 1$. In a point process we have, not a single non-negative integer-valued random variable, but a whole family of such variables. The probability generating function can, therefore, be generalized to a probability generating functional $G_N[\xi]$ which has similar properties and is defined by the equation

$$G_N[\xi] = E\left[\exp\left\{ \int_{\mathbb{R}} \log \xi(t) dN(t) \right\} \right]$$
$$= E\left\{ \prod_n \xi(T_n) \right\}, \qquad (2.41)$$

where $\{T_n\}$ are the (random) co-ordinates of the points. Where no ambiguity arises, the suffix N will be omitted. The two forms for $G[\xi]$ given in (2.41) are equivalent because, since N is a step function,

$$\int_{\mathbb{R}} \log \xi(t) dN(t) = \sum_n \log \xi(T_n).$$

Here $\exp\left\{ \int_{\mathbb{R}} \log \xi(t) dN(t) \right\}$ is defined to be zero if $\xi(T_n) = 0$ for some n and to be unity if $\xi(T_n) = 1$ for all n. The argument of $G[.]$, that is, the function ξ, must belong to some suitable class. We shall assume, in addition to technical regularity conditions, that $0 \le \xi(t) \le 1$ for all $t \in \mathbb{R}$ and that ξ is identically one outside some bounded set. This assumption ensures that in the product in (2.41) only a finite number of terms are not unity and, therefore, that the product converges.

As an example, the probability generating functional for a Poisson process with rate function $\rho(t)$ is

$$G[\xi] = \exp\left[-\int_{\mathbb{R}} \{1 - \xi(t)\} \rho(t) dt \right], \qquad (2.42)$$

which is similar to the probability generating function for a Poisson variable with parameter ρ, namely $\exp\{-\rho(1-z)\}$. Equation (2.42) may be proved as follows. Since $\xi(t)$ is identically one outside some bounded interval, we can denote this interval by I. Then $N(I)$ has a Poisson distribution with mean $P(I) = \int_I \rho(t) dt$, and given $N(I) = n$, the n points are all independently and identically distributed with density $\rho(t)/P(I)$ over I; see Section 3.1(i). Thus

$$
E\left[\exp\left\{\int_{\mathbb{R}} \log \xi(t)\,dN(t)\right\}\right]
$$

$$
= \sum_{n=0}^{\infty} \frac{e^{-P(I)}\{P(I)\}^n}{n!} \int_I dt_1 \ldots \int_I dt_n \frac{\xi(t_1)\rho(t_1)\ldots\xi(t_n)\rho(t_n)}{\{P(I)\}^n}
$$

$$
= e^{-P(I)} \sum_{n=0}^{\infty} \frac{1}{n!} \left\{\int_I \xi(t)\rho(t)\,dt\right\}^n
$$

$$
= \exp\left[-\int_I \{1 - \xi(t)\}\rho(t)\,dt\right]. \tag{2.43}
$$

Finally the integral over I in (2.43) may be extended to $\mathbb{R}$ since $1 - \xi$ vanishes outside I.

Analogously to the probability generating function, the generating functional provides a complete description of a unique point process via its finite dimensional distributions. Some simple properties of the process are straightforward to derive from the functional. For example, suppose that

$$
\xi(t) = \begin{cases} z & t \in A, \\ 1 & \text{otherwise,} \end{cases}
$$

where A is some bounded set. Then it follows immediately from (2.41) that

$$
G_N[\xi] = E\{z^{N(A)}\}, \tag{2.44}
$$

i.e. that the generating functional for the process N reduces to the generating function for $N(A)$. Similarly, if $A_1, \ldots, A_k$ are a sequence of disjoint sets and

$$
\xi(t) = \begin{cases} z_i & t \in A_i;\ i = 1, \ldots, k, \\ 1 & \text{otherwise,} \end{cases}
$$

then $G_N[\xi]$ reduces to the joint probability generating function of $N(A_1), \ldots, N(A_k)$.

Indeed $G_N[\xi]$ is essentially a convenient device for capturing the joint probability generating function for all possible choices of $A_1, \ldots, A_k$ and k. It is particularly important in obtaining theoretical results, for example limit theorems.

Another useful property of the probability generating function is that the generating function corresponding to a sum $Y_1 + Y_2$ of two independent variables Y_1 and Y_2, is the product of those corresponding to each variable separately. Therefore, whereas the distribution of a sum of k independent variables involves a $(k-1)$-fold convolution, the generating function of the sum is a k-fold product and is generally much easier to work with. This property is shared by generating functionals. The sum N of two independent processes N_1 and N_2 is termed the superposition and is such that $N(A) = N_1(A) + N_2(A)$ for all sets A. Then the generating functionals satisfy the relation

$$G_N[\xi] = G_{N_1}[\xi] G_{N_2}[\xi]. \tag{2.45}$$

For dealing with more general, not necessarily integer-valued, random measures one may define a Laplace functional and a characteristic functional by analogy with their counterparts for single-valued random variables.

2.8 Multivariate and multidimensional processes

Throughout this chapter it has been assumed that the point processes under discussion are univariate processes in one dimension, i.e. points all of the same class or type on the real line. For the most part the general theory of counts as described carries across to multivariate and multidimensional processes with only obvious modifications. Of course the interval specification of a process applies only to one-dimensional processes. For multivariate processes the counting measure N is taken to be vector-valued rather than single-valued. The various intensity functions, e.g. the intensity and the complete and conditional intensities, of the process must be defined for each type of point while, in addition, for the conditional intensity the type of point at the origin must also be specified. This applies also to the Palm distributions and the Palm measure. We shall develop these aspects further in Chapter 5.

Bibliographic notes, 2

The survey paper of Daley and Vere–Jones (1972) covers much of the theoretical content of this chapter in greater depth than has been attempted here and contains many historical references. Gnedenko and Kovalenko (1968) give a more elementary account. Orderliness is discussed in detail by Daley (1974). The existence of the occurrence parameter was proved by Khintchine (1960), who also proved the existence of the Palm distribution, derived the Palm–Khintchine equations and proved Korolyuk's theorem; for a more recent discussion of these topics, see Leadbetter (1968, 1972a). An early discussion of the properties of the sequence of intervals following an arbitrarily chosen time origin, is given by McFadden (1962).

Accounts of the general theory are given by Kallenberg (1976) and by Matthes, Kerstan and Mecke (1978). Some recent work has emphasized the connection with martingales (Brémaud and Jacod, 1977; Jacod, 1975; Liptser and Shiryayev, 1978, Chapter 18). For the existence of the complete intensity function, see Kallenberg (1976) and Papangelou (1974). Point processes arise in control theory, where they are often called jump processes; see Boel, Varaiya and Wong (1975) and Davis (1976).

In the text, second-moment properties and the complete intensity function have been strongly emphasized. In general discussion, higher-order moments and higher-order product densities are important (Bartlett, 1978, Section 3.42; Brillinger, 1978a; Krickeberg, 1974; Srinivasan, 1974).

The role of the spectrum in point processes was stressed by Bartlett (1963, 1964). A spectral representation for the point process itself, rather than for the associated covariance functions is possible (Daley and Vere–Jones, 1972; Doob, 1953, Section 11.11).

Early use of the probability generating functional in point processes was made by Moyal (1962); Westcott (1972) surveys that topic.

Further results and exercises, 2

2.1. Produce examples of point processes that are simply stationary but not stationary and which are weakly stationary but not simply stationary. [Section 2.2.]

2.2. Verify that a stationary or equilibrium renewal process with interval density g can be constructed as follows. Put points at $-u$ and v $(u, v \geq 0)$ with joint density

$$g(u + v) \bigg/ \int_0^\infty xg(x)dx.$$

Then put infinitely many points to the right of v and infinitely many to the left of $-u$, all with independent spacing governed by density g. [Section 2.2; Ryll–Nardzewski (1961).]

2.3. Suppose that a point process has a stationary sequence of intervals X_i such that

$$E(X_i) = \mu_X, \operatorname{cov}(X_i, X_{i+k}) = c_k \quad (k = 0, 1, \ldots).$$

Define $Y_i(i = 1, 2, \ldots)$ to be the interval between the ith and $(i + 1)$st points following a randomly chosen origin.

Show that

$$E(Y_i) = \mu_X + c_i/\mu_X$$

and hence that $n^{-1}(Y_1 + \ldots + Y_n)$ is a biased estimator of μ_X with bias

$$(n\mu_X)^{-1}(c_1 + \ldots + c_n).$$

[Sections 2.2, 1.2(ii); Cox and Lewis (1966) Chapter 4.]

2.4. Give an example of stationary point process of infinite rate, for example by taking a Poisson process whose rate is a random variable with a suitable distribution. Describe the realizations of the process. [Section 2.3.]

2.5. Use the Palm–Khintchine equations, for example in their generating function form, to prove that if $N(x)$ is the number of points in $(0, x]$ and $N_0(x)$ is the number of points in $(0, x]$ given a point at the origin, then $E\{N(x)\} = \rho x$ and

$$\text{var}\,\{N(x)\} = 2\rho \int_0^x [E\{N_0(u)\} - \rho u + \tfrac{1}{2}]\,du.$$

[Section 2.4; Cox (1962) p. 57.]

2.6. Obtain the asymptotic behaviour of var $\{N(t)\}$ as $t \to \infty$ for an ordinary renewal process by expanding about the origin the Laplace transform of (2.27). [Sections 2.5, 3.2(i); Cox (1962) Section 4.5.]

2.7. Write down the product density $p_2(t_1, t_2)$ for a stationary orderly point process with rate ρ and conditional intensity function h. Hence deduce (2.27). [Section 2.5.]

2.8. Use (2.42) to deduce the distribution of the time to the first point after the origin, in a Poisson process with rate $\rho(t)$, by considering the probability of no points occurring in some interval $(0, u]$. [Sections 2.7, 3.1(ii).]

2.9. Derive the probability generating functional for a Poisson process with rate $\rho(t)$, given in (2.42), by first approximating $\int \log \xi(t)\,dN(t)$ by the sum $\Sigma \log \xi(t_i) \Delta N(t_i)$, where $\Delta N(t_i) = N(t_i, t_{i+1})$ and $\{t_i\}$ defines a partition of the real line with $t_{i+1} - t_i = \delta$ for all i, and then letting $\delta \to 0$. [Section 2.7.]

2.10. A point process N has the probability generating functional

$$G[\xi] = \exp\Bigg[- \int_{-\infty}^{\infty} \{1 - \xi(t)\} \rho(t) dt$$

$$+ \frac{1}{2} \int_{-\infty}^{\infty} \int_{-\infty}^{\infty} \{1 - \xi(t)\} \{1 - \xi(u)\} q(t, u) dt du \Bigg],$$

where $q(t, u) = q(u, t)$. Show that if A and B are disjoint bounded sets then

$$\operatorname{cov} \{N(A), N(B)\} = \int_A dt \int_B du\, q(t, u).$$

[Sections 2.7, 4.6.]

2.11. A point process is constructed from a homogeneous Poisson process of rate ρ by adding extra points. Corresponding to each point of the Poisson process a new point is added independently so that if a Poisson process point is at t then a new point is added at $t + \tau$ with probability density $f(\tau)$, $\tau \in \mathbb{R}$. Show that the process of all the points has probability generating functional

$$\exp\Bigg[- \rho \int_{-\infty}^{\infty} dt \Big\{ 1 - \int_{-\infty}^{\infty} d\tau\, f(\tau) \xi(t) \xi(t + \tau) \Big\} \Bigg].$$

[Sections 2.7, 3.4.]

CHAPTER 3

Special models

3.1 Poisson processes

(i) *Introduction*

The homogeneous Poisson process is, in many ways, the simplest point process, and it will be seen later that it plays a role in point process theory in most respects analogous to that of the normal distribution in the study of random variables. The Poisson process has been defined in Section 1.2(i) as an orderly process for which the complete intensity function is a constant, ρ. From this definition the counting properties of the process are easily derived. In particular the process has the property C1 as follows:

C1. For any $k = 1, 2, \ldots$ the numbers of points $N(A_1), \ldots, N(A_k)$ in arbitrary disjoint sets $A_1, \ldots, A_k$ have independent Poisson distributions with means $\rho|A_1|, \ldots, \rho|A_k|$.

The interval properties of the Poisson process are also easily deduced from our definition. If 0 is an arbitrary time origin and the process points are labelled according to the convention $\ldots \le T_{-1} \le T_0 \le 0 < T_1 \le \ldots$, then the process has the following property C2:

C2. The intervals $\ldots, T_{-1} - T_{-2}, T_0 - T_{-1}, -T_0, T_1, T_2 - T_1, \ldots$ are a sequence of independent exponentially distributed variables all with the same parameter ρ.

Clearly, for a Poisson process, the Palm probability $\pi_k(x)$, defined in Section 2.4 as the limiting probability of k points in the interval $(0, x]$ given a point near 0, is equal to the unconditional probability $p_k(x)$ of k points in the interval $(0, x]$.

Alternative definitions of the Poisson process are possible; in particular each of the properties C1 and C2 is sufficient to characterize the process. It is, however, easy to construct a non-Poisson point process with dependent intervals for which the marginal distributions are, nevertheless, the exponential distribution with parameter ρ. Similarly, one may construct a process for which the number of points in any interval I has a Poisson distribution with mean $\rho|I|$, but which is not a Poisson process, because numbers of points in disjoint sets are not independent. An example is given in Section 3.2(iii).

Many other characterizations of the Poisson process are possible. For example, if a renewal process satisfies any one of C3, C4 or C5, then it must be a Poisson process:

C3. The expected forward recurrence time from t does not depend on t, for all $t > 0$ (Çinlar and Jagers, 1973);

C4. For some $t > 0$, the distribution of the forward recurrence time from t is the same as the interval distribution of the renewal process, where this latter distribution is non-arithmetic, i.e. is not concentrated on any set $\{0, a, 2a, \ldots\}$ with $a > 0$ (Isham, Shanbhag and Westcott, 1975);

C5. The backward and forward recurrence times from t are independent, for some $t > 0$ (Erickson and Guess, 1973).

Another important property of the Poisson process is that, given that n points occur in an interval, those points are independently and uniformly distributed over the interval. For, consider without loss of generality the interval $(0, t]$; then if $0 = t_0 < t_1 < \ldots < t_n \leq t$,

$$\lim_{\delta_1, \ldots, \delta_n \to 0+} (\delta_1 \ldots \delta_n)^{-1} \operatorname{pr} \{N(t) = n, N(t_i, t_i + \delta_i) = 1; i = 1, \ldots, n\}$$

$$= \left\{ \prod_{i=1}^{n} \rho e^{-\rho(t_i - t_{i-1})} \right\} e^{-\rho(t - t_n)}$$

$$= \rho^n e^{-\rho t},$$

so that

$$\lim_{\delta_1, \ldots, \delta_n \to 0+} (\delta_1 \ldots \delta_n)^{-1} \operatorname{pr} \{N(t_i, t_i + \delta_i) = 1; i = 1, \ldots, n | N(t) = n\}$$

$$= n!/t^n, \tag{3.1}$$

which is the joint probability density for the order statistics of n independent variables each uniformly distributed over an interval of length t. In the same way it can be shown that if the Poisson process is non-homogeneous with rate $\rho(t)$ then, given that there are n points in the interval I, these points are independent and identically distributed with density $\rho(t)/P(I)$, where $P(t) = \int_0^t \rho(u) du$.

Although the points of a homogeneous Poisson process, given the total number of points in an interval, are independently and uniformly distributed over the interval, nevertheless realizations of the process tend to exhibit apparent clustering. This can be seen in the realization in Fig. 1.3(i). There are various ways in which this effect can be formalized. For example, one might look at the largest number of points covered by an interval of fixed length as it moves along the realization, or alternatively perhaps, find the size of the smallest interval containing a pre-assigned number of points. Suppose that ten points of a Poisson process lie in the interval $[0, 1]$ and let U be the largest number of these points which lie in a sub-interval of length 0.1. Then it can be shown that $E(U) = 3.3$, whereas the expected number of points in a random interval of this length is one (Naus, 1966).

The Poisson process is of fundamental importance in a variety of limiting situations which will be mentioned only briefly here and discussed in more detail later. In the theory of point processes, the analogue of a sum of random variables is a superposition of point processes; if k individual processes have counting measures $N_1, \ldots, N_k$, then their superposition has counting measure N satisfying $N = N_1 + \ldots + N_k$. For a particular set A, if the variables $N_i(A)$ $(i = 1, \ldots, k)$ are well-behaved, the Central Limit Theorem applies and $N(A)$ is approximately normally distributed for large k. Further, we can study the behaviour of the superposition N as a point process. Then the result for point processes, which is in a sense analogous to the Central Limit Theorem for random variables, is that, if the individual processes are well-behaved, their superposition is asymptotically a Poisson process. Note that since, if the mean is large, the Poisson distribution is approximately normal, there is no contradiction here with respect to the distribution of $N(A)$. Intuitively, 'well-behaved' means that the individual processes must be sufficiently sparse, with no single process dominating the rest. This result will be discussed more fully in Section 4.5; it explains why many physical processes can be approximated closely by a Poisson process, since they can be described in terms of a superposition. Calls from many subscribers arriving at a telephone exchange form one important example.

Another characterization of the Poisson process is linked with superpositions, namely, the superposition of two independent renewal processes is a renewal process if and only if the components are Poisson processes (Samuels, 1974).

The Poisson process also arises in other limiting situations; again detailed discussion of these is postponed to Chapter 4. Briefly, under certain conditions, if the points of a process are independently subjected to displacements which are identically distributed according to a sufficiently dispersed distribution, then the resulting process is asymptotically Poisson. Also, if the points of a process are independently deleted with probability $1 - p$ and retained with probability p, and if the retained points are suitably rescaled, then the resulting process tends to a Poisson process as p tends to zero.

Thus the Poisson process occurs in many limiting situations. It also has the property of complete randomness in two senses: its intensity is constant irrespective of the history of the process, and, given that $N(t) = n$, the n points are independently and uniformly located over $(0, t]$. For these reasons the Poisson process is of central importance among point processes, both in theory and in practice. The Poisson process also provides a natural starting point for the construction of processes with more complex structure and indeed many important special processes are generalizations of it. Some of these will be described only briefly here and will be discussed in more detail in their own right in subsequent sections of this chapter.

(ii) *Non-stationary Poisson processes*

The simplest generalization of the homogeneous Poisson process is the non-stationary Poisson process in which the rate is a function $\rho(t)$ of time. This of course includes the situation in which the rate is a function of some observed explanatory variable $z(t)$. The probability generating functional for this process has already been derived in Section 2.7 and is

$$G[\xi] = \exp\left[- \int_{-\infty}^{\infty} \{1 - \xi(t)\} \rho(t) dt \right], \qquad (3.2)$$

from which all properties of the process may, in principle, be recovered. Alternatively, a simple transformation of the time scale reduces the non-homogeneous process to a homogeneous one. If a new time variable τ, sometimes called operational time, is defined by

$$\tau(t) = \int_{0}^{t} \rho(u) du,$$

so that τ is identified as the mean number of points in $(0, t]$, then on the new time scale we have a Poisson process with unit rate. For, clearly, the change in time scale will not affect the property that counts in disjoint sets are independent and, in terms of the transformed time τ, the process has unit rate.

The important features of the process are that if $A_1, \ldots, A_k$ are arbitrary disjoint sets then $N(A_1), \ldots, N(A_k)$ are still independent Poisson variables but with

$$E\{N(A_i)\} = \int_{A_i} \rho(t) dt \quad (i = 1, \ldots, k).$$

Now the intervals between points of the process are not independent;

further the non-stationarity of the process complicates their distributions. For example, an interval X which starts from a point at t_0 has density

$$f_X(x; t_0) = \rho(t_0 + x) \exp \left\{ - \int_{t_0}^{t_0 + x} \rho(t) dt \right\}, \qquad (3.3)$$

with mean

$$E(X; t_0) = \int_0^\infty dx \exp \left\{ - \int_{t_0}^{t_0 + x} \rho(t) dt \right\}.$$

To see the effect of the non-constant rate function $\rho(t)$, consider the behaviour of the survivor function $\mathscr{F}_X$, where without loss of generality we set $t_0 = 0$, so that

$$\mathscr{F}_X(x) = \exp \left\{ - \int_0^x \rho(t) dt \right\}.$$

We assume that ρ is effectively linear over the range of x of interest with

$$\rho(x) = \rho_0(1 + \varepsilon \rho_0 x) + o(\varepsilon),$$

where ε is small and dimensionless. Then

$$\mathscr{F}_X(x) = \exp \left\{ - \left(\rho_0 x + \frac{1}{2} \varepsilon \rho_0^2 x^2 \right) \right\} + o(\varepsilon)$$

$$= \left\{ 1 + \varepsilon \left(x \rho_0 - \frac{1}{2} x^2 \rho_0^2 \right) \right\} \exp \left\{ - \rho_0(1 + \varepsilon) x \right\} + o(\varepsilon), (3.4)$$

where in (3.4) the survivor function is expressed so that the correction term makes no contribution to the mean of the distribution. Thus to first order the effect of the time dependence is to change the parameter of the exponential to $(1 + \varepsilon) \rho_0$ and to perturb the survivor function by a term with the sign of $\rho'(0)$ for $x < 2/\rho_0$ and with the opposite sign for $x > 2/\rho_0$.

(iii)   *Compound Poisson processes*

As mentioned in Section 1.4(i), another simple generalization of the Poisson process, this time to a process with multiple occurrences, is the compound Poisson process. In this, each point of the Poisson process is replaced by a random number M of simultaneous occurrences, the multiplicities corresponding to distinct points of the

Poisson process being independent and identically distributed. The probability generating functional for this process is easily shown to be

$$G[\xi] = \exp\left\{ -\rho \int_{-\infty}^{\infty} [1 - G_M\{\xi(t)\}]\,dt \right\}, \qquad (3.5)$$

where $G_M(z) = E(z^M)$ is the probability generating function for M, since we need only substitute $G_M\{\xi(t)\}$ as the argument in the generating functional of the homogeneous Poisson process. Properties of the compound Poisson process can be obtained from (3.5), but in most cases it will be simplest to start from first principles using the construction of the process in terms of a Poisson process of instants of occurrence, together with the multiplicities of the occurrences. The mean number of occurrences in a set A is clearly $E(M)\rho|A|$ and the numbers in disjoint sets are independent. The intervals between instants of occurrences are still independent exponential variables.

(iv) Cluster processes

In the compound Poisson process of Section 3.1(iii) each point of the Poisson process is replaced by a cluster of points, where the original point is regarded as the cluster centre. The cluster sizes are independent and identically distributed and the cluster points are all located at the cluster centre. It is usually convenient to assume that the cluster centre is not observed. This involves no loss of generality, for otherwise we take $\text{pr}(M = 0) = 0$. More general cluster processes in which the cluster points have some distribution of location relative to their cluster centre are considered in Section 3.4. It is usually simplest to retain the property of independent, identically distributed cluster sizes and a Poisson process of cluster centres, although these assumptions may be weakened. Two specific Poisson cluster processes are particularly tractable, the Neyman–Scott cluster process in which the points in the cluster are independently and identically distributed about the cluster centre, and the Bartlett–Lewis cluster process in which the cluster points follow the cluster centre in a finite renewal process.

(v) Doubly stochastic Poisson processes

In the non-homogeneous Poisson process, the rate function is a known function of time. However, the Poisson process may also be generalized by using an unobserved stochastic process, $\{\Lambda(t)\}$, for the rate function, in which case the resulting process is a doubly stochastic Poisson process, as already introduced in Section 1.2(iv). Such processes will be discussed in more detail in Section 3.3(iii).

3.2   Renewal processes and generalizations

(i)   *Introduction*

A point process may be specified either in terms of counts of points in sets or in terms of its interval sequence, as outlined in Section 2.1. For a particular process one of these approaches is usually simpler than the other, although we have seen that for the Poisson process both specifications are straightforward. In this section we shall concentrate on processes which have simple specifications in terms of the sequence of intervals.

For the Poisson process, the intervals between points are independent exponential variables, so that a natural generalization is to a process in which the intervals are independent with some common distribution function G. Such processes are known as renewal processes and have already been described briefly in Section 1.2(ii). The properties of renewal processes are well-documented, for example by Cox (1962), so that only some of the more important aspects will be described here.

In defining a renewal process it is convenient to specify the process on the positive real line in terms of the interval X_1 from the origin to the first point of the process after the origin and the intervals X_i $(i = 2, 3, \ldots)$ between the $(i-1)$st and ith points following the origin. The intervals $X_1, X_2, \ldots$ are taken to be a sequence of mutually independent variables. When $X_1, X_2, \ldots$ all have the same distribution function G, the renewal process is said to be ordinary, and effectively the process is defined as if a point occurs at the origin. The renewal process can be extended in an obvious way to the whole real line by defining further independent variables $X_{-1}, X_{-2}, \ldots$, with the same distribution function G, to be the intervals between points occurring before the origin.

If the first interval X_1 has a different distribution G_1 from subsequent intervals $X_2, X_3, \ldots$, which still have the same distribution function G, then the process is a modified renewal process. The implicit assumption is thus that the process does not start from a point at the origin but, usually, from part way through an interval. This makes particular sense if $X_2, X_3, \ldots$ are lifetimes of identical new components, but X_1 is the residual lifetime of a component which has already been in use. Again the process can be extended to the negative real line if required. As already mentioned in Section 1.2(ii), a particular distribution is sometimes assumed for X_1, namely that X_1 has density function $g_1(x) = \mu_X^{-1} \mathscr{G}(x)$, where μ_X is the mean of the distribution G and $\mathscr{G}$ is the survivor function. This process is termed an equilibrium renewal process and is, as the name implies, a stationary process in the sense of Section 2.2. Since the only distribution G with density g which satisfies

$$g(x) = \mu_X^{-1} \mathscr{G}(x)$$

is the exponential distribution with $g(x) = \mu_X^{-1} \exp(-x/\mu_X)$, the ordinary renewal process is a stationary process if and only if it is a Poisson process. Note that we define a renewal process to have no multiple occurrences, that is G_1 and G do not have atoms at the origin.

The properties of renewal processes to be described here will be given for ordinary renewal processes; changes to cover modified renewal processes are easily made. Since a renewal process is defined in terms of a sequence of independent intervals, properties based on intervals are particularly simple. Thus, the time $S_k = X_1 + \ldots + X_k$ from the origin to the kth point of the process has distribution function $G^{(k)}$, the k-fold convolution of G. The counting properties are then obtained from the interval properties by using equivalences of the form

$$N(t) < k \quad \text{if and only if} \quad S_k > t,$$

which leads to

$$\mathrm{pr}\{N(t) = k\} = G^{(k)}(t) - G^{(k+1)}(t). \tag{3.6}$$

The mean of the distribution of $N(t)$ given in (3.6) is called the renewal function $H(t)$, and satisfies

$$H(t) = E\{N(t)\} = \sum_{k=1}^{\infty} k\{G^{(k)}(t) - G^{(k+1)}(t)\}$$

$$= \sum_{k=1}^{\infty} G^{(k)}(t). \tag{3.7}$$

The theoretical properties of renewal processes, since they involve convolutions of distributions, are often most concisely written in terms of Laplace transforms. In particular, the transform of (3.7) is

$$H^*(s) = \int_0^{\infty} H(t) e^{-st} dt = \frac{1}{s} \sum_{k=1}^{\infty} \{g^*(s)\}^k$$

$$= \frac{1}{s} \frac{g^*(s)}{1 - g^*(s)}, \tag{3.8}$$

where

$$g^*(s) = \int_0^\infty e^{-st} g(t) dt$$

and g is the density function corresponding to G, regarded if necessary as a generalized function.

The distribution of S_k is, by the Central Limit Theorem, approximately normal for large k and we have shown in Section 2.5 that this implies the asymptotic normality of $N(t)$ as $t \to \infty$, with

$$H(t) = E\{N(t)\} \sim t/\mu_X \tag{3.9}$$

and

$$\text{var}\{N(t)\} \sim \sigma_X^2 t/\mu_X^3, \tag{3.10}$$

where σ_X^2 is the variance of the interval distribution G of the renewal process.

Differentiating (3.7) we obtain

$$h(t) = \frac{d}{dt} H(t) = \sum_{k=1}^\infty g^{(k)}(t) \tag{3.11}$$

and directly, or from (3.8),

$$h^*(s) = \frac{g^*(s)}{1 - g^*(s)},$$

from which it follows, on expansion as $s \to 0$, that $h(t) \to 1/\mu_X$ as $t \to \infty$. The function h is known in this context as the renewal density. However, since it refers to a process with a point at the origin, it is the same function h that was defined in (1.19) as the conditional intensity and can be interpreted for the ordinary renewal process as

$$h(t) = \lim_{\delta \to 0+} \delta^{-1} \text{pr}\{N(t, t+\delta) > 0\}.$$

Then, if V_t is the forward recurrence time from an arbitrary time t,

$$\text{pr}(V_t > x) = \mathcal{G}(t+x) + \int_0^t h(t-u)\mathcal{G}(u+x) du. \tag{3.12}$$

For, either
   (i) the next point after t is the first point of the process, or
   (ii) the last point to occur at or before t is at $t - u$ and is followed by an interval of length at least $u + x$, for some $u \in [0, t)$.

Since $h(t) \to 1/\mu_X$ as $t \to \infty$, it follows from (3.12) that

$$\mathrm{pr}(V_t > x) \to \mu_X^{-1} \int_x^\infty \mathscr{G}(u)\,du;$$

this is an alternative proof of the result of (1.9) that the forward recurrence time from an instant chosen uniformly over a long time interval will have a density which approaches $\mathscr{G}(x)/\mu_X$ as the length of that time interval tends to infinity. It also follows that if X_1 has density $\mathscr{G}(x)/\mu_X$, then the renewal process on $(0, \infty)$ behaves as if it were an ordinary renewal process with an origin at $-\infty$, i.e. is in equilibrium.

The intervals of the Poisson process are independent exponential variables, while in a renewal process the intervals are independent and identically distributed. One way of weakening these assumptions further is to retain the independence of the intervals while allowing different intervals to have different distribution functions. A 'time-dependent' renewal process can be defined by assuming that the nth interval X_n has distribution function G_n, a known function of n. The properties of the renewal process can be extended in a straightforward way, in theory at least, to cover this case. Since the distribution of $S_k = X_1 + \ldots + X_k$ is the multiple convolution of $G_1, \ldots, G_k$ these properties will have a simple algebraic form if the same is true of this convolution, or its Laplace transform.

We now consider some other ways of generalizing the renewal process via its interval sequence.

(ii) Semi-Markov processes
In the 'time-dependent' renewal process mentioned above, the sequence of interval distributions is known. A convenient way of specifying a stochastic sequence of distribution functions for the interval sequence of the process is to use a Markov chain. For example, suppose that we have k distribution functions $F_1, \ldots, F_k$ and that we say the process is in 'state' j at a particular time if the current distribution function is F_j. Then we assume that the sequence of states of the process is determined by a Markov chain with a matrix $P = ((p_{ij}))$ of transition probabilities, so that if the process is in state i the probability is p_{ij} that the next state will be state j. The point process determined by a sequence of intervals of this sort is called a semi-Markov process. Some authors call this point process a Markov renewal process, reserving the term semi-Markov process for the state as a function of time.

An alternative way of describing this process is to assume that the process points each have a class or type; that is, a random variable taking one of the values $1, \ldots, k$ is associated with each point and the sequence of types is determined by a Markov chain with transition matrix P. Then for each $i = 1, \ldots, k$ the intervals which start with a class i point have distribution function F_i, all the intervals being mutually independent. A more general semi-Markov process has a sequence of point classes determined as before, but then the interval which starts with a class i point and ends with a class j point is taken to have distribution function F_{ij}, where now we have k^2 distribution functions F_{ij} $(i, j = 1, \ldots, k)$. The semi-Markov process can also be regarded as a generalization of a Markov process in which the time spent in a particular state between transitions is no longer exponentially distributed.

As with the renewal process we assume that $F_i(0) = 0$, $F_{ij}(0) = 0$ $(i, j = 1, \ldots, k)$, so that the semi-Markov process has no multiple occurrences.

As an example, a semi-Markov process with just two classes of point, has a transition matrix P of the form

$$P = \begin{pmatrix} p_1 & 1 - p_1 \\ 1 - p_2 & p_2 \end{pmatrix}, \tag{3.13}$$

and four interval distributions F_{11}, F_{12}, F_{21} and F_{22}. In the simpler type of process, in which the interval distribution depends only on the type of the point at the beginning of the interval, we take $F_{11} \equiv F_{12} \equiv F_1$, and $F_{21} \equiv F_{22} \equiv F_2$ say. A special case of these semi-Markov processes is the alternating renewal process. Here $p_1 = p_2 = 0$, so that the intervals of the point process are independent with distributions which are alternately F_1 and F_2. The alternating renewal process might, for example, be a suitable model for a machine alternately working and under repair. It would be required that the working times all have the same distribution, that the repair times all have the same distribution and that all these intervals are independent. In particular, the repair time following a break-down must not depend on the length of time since the previous repair.

In any semi-Markov process for which $F_{ij} \equiv F_i$ $(i, j = 1, \ldots, k)$ the number of consecutive intervals which have the same distribution is geometrically distributed. Thus a generalization of the semi-Markov process, when $k = 2$, which has been suggested for modelling spike trains recorded from nerve cells, is a process in which the numbers of identically distributed intervals follow non-geometric distributions (Ekholm, 1972). This process has random numbers of independent intervals from the distribution F_1 alternating with random numbers of independent intervals from the distribution F_2, where the random numbers are themselves mutually independent with one distribution for the numbers of type 1 intervals and a second distribution for the numbers of type 2 intervals.

Many of the properties of semi-Markov processes are straightforward to derive from those of renewal processes. For example, the points of a particular type form a renewal process, so that if it is these points which are of interest, then it is necessary to consider only the distribution of an interval between successive points of this type and to make use of the standard results of renewal theory. For the alternating renewal process this interval distribution is simply the convolution of distributions F_1 and F_2.

The semi-Markov process is, as defined, a multivariate process and its properties can be investigated by the methods of Chapter 5. However, the situation may arise in which the classes or types of the points are not observed. Then, although the underlying structure is that of a semi-Markov process, the observed process is univariate. If one regards the semi-Markov process as consisting of k dependent processes $N^{(1)}, \ldots, N^{(k)}$, where $N^{(i)}$ refers to the points of class i, the observed process of points is the superposition $N = N^{(1)} + \ldots + N^{(k)}$.

Suppose that the Markov chain with transition matrix P has an equilibrium distribution given by the row vector π, where π satisfies $\pi = \pi P$. Then, if we assume that the semi-Markov process is in equilibrium, the distribution F_X of an interval X between points in the observed univariate process satisfies

$$F_X(x) = \sum_{i,\,j=1}^{k} \pi_i p_{ij} F_{ij}(x), \qquad (3.14)$$

with mean

$$E(X) = \sum_{i,\,j=1}^{k} \pi_i p_{ij} \mu_{ij}$$

and variance

$$\mathrm{var}(X) = \sum_{i,\,j=1}^{k} \pi_i p_{ij} (\sigma_{ij}^2 + \mu_{ij}^2) - \left(\sum_{i,\,j=1}^{k} \pi_i p_{ij} \mu_{ij} \right)^2,$$

where

$$\int_0^\infty x f_{ij}(x)\,dx = \mu_{ij}, \qquad \int_0^\infty (x - \mu_{ij})^2 f_{ij}(x)\,dx = \sigma_{ij}^2,$$

and $f_{ij}(x)$ is the density function corresponding to $F_{ij}(x)$, regarded if necessary as a generalized function.

It is straightforward to write down the second-order properties of
the interval sequence. For example, if we want the autocovariance
between intervals of lag $r + 1$, then we must condition on the types of
points at the ends of each interval; these intervals will have
independent distributions F_{ij} and F_{lm} with probability $\pi_i p_{ij} p_{jl}^{(r)} p_{lm}$,
where $p_{jl}^{(r)}$ is the r-step transition probability between states j and l,
that is the (j, l)th element of P^r.

The second-order properties of counts for the observed univariate
process are also straightforward, though notationally rather tedious.
It is convenient to consider the conditional cross-intensity functions
$h_i^{(j)}$ $(i, j = 1, \ldots, k)$ defined by

$$h_i^{(j)}(t) = \lim_{\delta_1, \delta_2 \to 0+} \delta_2^{-1} \mathrm{pr}\{N^{(j)}(t, t + \delta_2) > 0 \,|\, N^{(i)}(-\delta_1, 0) > 0\}.$$

Then the $h_i^{(j)}$ must satisfy the set of k^2 equations

$$h_i^{(j)}(t) = p_{ij} f_{ij}(t) + \sum_{l=1}^{k} \int_0^t h_i^{(l)}(t - u) p_{lj} f_{lj}(u) \, du \qquad (3.15)$$

$$(i, j = 1, \ldots, k).$$

For, either
    (i) there are no process points in $(0, t]$ or
    (ii) the last point before the type j point at $t+$ has type l, for some l,
and occurs at $t - u$ for some u.

The Laplace transform of (3.15) is, for $i, j = 1, \ldots, k$,

$$h_i^{(j)*}(s) = \int_0^\infty e^{-st} h_i^{(j)}(t) \, dt = p_{ij} f_{ij}^*(s) + \sum_{l=1}^{k} h_i^{(l)*}(s) p_{lj} f_{lj}^*(s),$$

$$(3.16)$$

where

$$f_{ij}^*(s) = \int_0^\infty e^{-st} f_{ij}(t) \, dt.$$

If we define matrices $h^*(s)$ and $\alpha^*(s)$ by

$$h^*(s) = ((h_i^{(j)*}(s))), \quad \alpha^*(s) = ((p_{ij} f_{ij}^*(s))),$$

then the set of equations (3.16) can be written in matrix form as

$$h^*(s)\{I - \alpha^*(s)\} = \alpha^*(s),$$

where I is the $k \times k$ identity matrix.

Now, if the semi-Markov process is in equilibrium, then the
conditional intensity function h for the observed univariate process is

given by

$$h(t) = \sum_{i,\,j=1}^{k} \pi_i h_i^{(j)}(t). \tag{3.17}$$

The conditional intensity function can then be used as described in Section 2.5 to obtain the second-order properties of the counts for the observed process. Note, though, that the function h has been obtained in terms of its Laplace transform and in general this will be difficult to invert. However, in some special cases inversion will be straightforward, for example, if the interval distributions are exponential.

(iii) *Moran's process with pairwise dependent intervals*

Another way of generalizing the sequence of independent and identically distributed intervals of a renewal process is to weaken the independence criterion, and one simple process of this sort is formed from independent and identically distributed pairs of dependent intervals, which have the same marginal distribution. Thus the interval sequence for the process is $\ldots, U_{i-1}, V_{i-1}, U_i, V_i, U_{i+1}, \ldots,$ where $\{(U_i, V_i)\}$ is a sequence of independent pairs (U_i, V_i) which each have some bivariate distribution F_{UV} with the property that the marginal distributions of U and V are the same. As with the renewal process, the origin may be specified at the end of either a U or a V interval or in the middle of an interval. The process is in equilibrium if the origin is chosen in a way similar to that used for an equilibrium renewal process. If $W_i = U_i + V_i$, then the W_i form a renewal process and the origin can be fixed in a particular interval W_0, so that the $\{W_i\}$ process is in equilibrium. It then remains to divide each W_i into intervals U_i, V_i using the conditional joint distribution of U_i and V_i given that $U_i + V_i = W_i$.

A particular example of this process (Moran, 1967) takes the joint density of U and V as

$$f(u, v) = e^{-(u+v)} + f_\varepsilon(u, v),$$

where $f_\varepsilon(u, v)$ takes values $\pm \varepsilon (0 < \varepsilon < e^{-6})$ in the regions indicated in Fig. 3.1 and is zero elsewhere. This rather artificial joint density is such that U and V are marginally exponentially distributed with unit mean and their sum has the same distribution as the sum of two independent exponential variables with unit mean, namely $W = U + V$ has density we^{-w}. Thus the sum of any two adjacent intervals of the process, whether or not they are dependent, has this same density. It can then be shown that the distribution of the number of points in

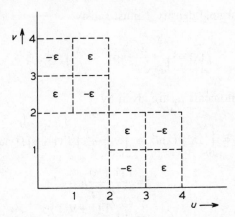

Fig. 3.1. *Definition of Moran's process. Joint density of pair of successive intervals e^{-u-v}, modified by $\pm \varepsilon$ as shown.*

any interval has a Poisson distribution with mean equal to the length of the interval. This process provides an example of the situation described in Section 3.1(i) in which a process has intervals which are marginally exponentially distributed, the number of points in any interval has a Poisson distribution and yet the process is not a Poisson process.

(iv)   *Processes with Markov-dependent intervals*
An intuitively appealing process with dependent intervals has the intervals $\{X_j\}$ forming a Markov sequence, so that the distribution of X_{j+1} given $\{X_i, i \le j\}$ does not depend on $\{X_i, i < j\}$. More generally, the intervals might be taken to form an m-dependent Markov sequence so that the distribution of X_{j+m} given $\{X_i, i \le j + m - 1\}$ does not depend on $\{X_i, i < j\}$. We consider some examples for $m = 1$. Then if the sequence of intervals is to be stationary it is sufficient to specify only the marginal interval density f and the conditional density $f(.; y)$ of X_{j+1} given $X_j = y$, which must not depend on j. These two densities must satisfy

$$f(x) = \int_0^\infty f(x; y) f(y) dy. \tag{3.18}$$

If we assume a particular form for $f(x; y)$ it is in general difficult to solve (3.18) for $f(x)$. Both Wold (1948, 1949) and Cox (1955) assumed that the conditional distribution of X_{j+1} given $X_j = y$ is exponential with a mean which depends on y. Wold took this mean to be $\sqrt{y}/\lambda$.

Then the marginal density f must satisfy

$$f(x) = \int_0^\infty \frac{\lambda}{\sqrt{y}} \exp\left(-\frac{\lambda x}{\sqrt{y}}\right) f(y) dy,$$

so that the moments μ_α are given by

$$\mu_\alpha = \int_0^\infty x^\alpha f(x) dx = \int_0^\infty \left(\frac{\sqrt{y}}{\lambda}\right)^\alpha \Gamma(1 + \alpha) f(y) dy$$

$$= \frac{\Gamma(1 + \alpha)}{\lambda^\alpha} \mu_{\alpha/2}$$

$$= \prod_{k=0}^\infty \frac{\Gamma(1 + \alpha/2^k)}{\lambda^{\alpha/2^k}}.$$

Another possibility is that the exponential distribution of X_{j+1} given $X_j = y$ has parameter $\lambda_0(1 + \varepsilon \lambda_0 y)$, where ε is small and dimensionless. Then (3.18) can be solved to give an approximate solution for the marginal density f of the form

$$f(x) = \lambda_0 e^{-\lambda_0 x}[1 + \varepsilon(1 - \lambda_0 x) + \varepsilon^2\{(\lambda_0 x)^2 - \lambda_0 x - 1\}] + o(\varepsilon^2)$$

$$= \lambda_0(1 + \varepsilon - 2\varepsilon^2)e^{-\lambda_0(1 + \varepsilon - 2\varepsilon^2)x}[1 + \varepsilon^2\{\tfrac{1}{2}(\lambda_0 x)^2 - 2\lambda_0 x + 1\}]$$

$$+ o(\varepsilon^2),$$

where the leading term is an exponential density with parameter $\lambda_0(1 + \varepsilon - 2\varepsilon^2)$ and the correction term, which has order ε^2, makes no contribution to the mean of the distribution. Note that the factor $\tfrac{1}{2}(\lambda_0 x)^2 - 2\lambda_0 x + 1$ is $L_2(\lambda_0 x)$, where L_2 is the second degree Laguerre polynomial. Note also that if $\varepsilon < 0$, the formal probability of a negative parameter value in the exponential distribution of X_{j+1} given X_j is $O(e^{1/\varepsilon})$ and can be ignored.

If a consistent pair of densities $f(x)$ and $f(x; y)$ are given, then the process is completely specified, but even so its properties are, in general, difficult to obtain. Suppose that we consider the number $N_y(y, t)$ of points in $(y, t]$ when the process has a point at y and let Y be the interval starting from a particular point and let Z be the interval immediately preceding that point. Then

$$\text{pr}\{N_0(0, t) = n\} = \int_0^\infty \text{pr}\{N_0(0, t) = n | Z = z\} f(z) dz \quad (3.19)$$

$$(n \geq 0),$$

where

$$\text{pr}\,\{N_0(0,t)=n\,|\,Z=z\} = \int_0^t \text{pr}\,\{N_y(y,t)=n-1\,|\,Y=y\}\,f(y;z)\,dy$$

$$= \int_0^t \text{pr}\,\{N_0(0,t-y)=n-1\,|\,Z=y\}\,f(y;z)\,dy$$

$$(n \geq 1),$$

with

$$\text{pr}\,\{N_0(0,t)=0\,|\,Z=z\} = \mathscr{F}(t;z) = \int_t^\infty f(y;z)\,dy.$$

Note that for $n=0$, equation (3.19) is, as expected,

$$\text{pr}\,\{N_0(0,t)=0\} = \int_0^\infty \mathscr{F}(t;z)\,f(z)\,dz$$

$$= \mathscr{F}(t),$$

where $\mathscr{F}(t)$ is the marginal survivor function for the intervals.

If the process is in equilibrium and $N(0,t)$ is the number of points in $(0,t]$, the time to the first point after the origin has the 'forward recurrence time' density $\mu_X^{-1}\mathscr{F}(x)$, where μ_X is the mean of the marginal interval distribution, so that

$$\text{pr}\,\{N(0,t)=0\} = \int_t^\infty \mu_X^{-1}\mathscr{F}(x)\,dx. \tag{3.20}$$

The joint density of forward and backward recurrence times from the origin is $\mu_X^{-1}f(x+y)$, and it follows that for $n \geq 1$

$$\text{pr}\,\{N(0,t)=n\}$$

$$= \int_0^\infty dy \int_0^t dx\,\mu_X^{-1}f(x+y)\,\text{pr}\,\{N_0(0,t-x)=n-1\,|\,Z=x+y\}. \tag{3.21}$$

Second-order properties both of intervals and of counts are more difficult. For the former we must find the density of X_{j+k} given $X_j = x$, $f_{(k)}(z;x)$ say, for $k=1,2,\ldots$, which satisfies

$$f_{(k)}(z;x) = \int_0^\infty f_{(k-1)}(z;y)\,f(y;x)\,dy \quad (k>1),$$

with

$$f_{(1)}(z; x) = f(z; x).$$

For the second-order properties of counts, we need to find the distribution of the sum $S_k = X_1 + \ldots + X_k$ for each $k = 1, 2, \ldots$, because the conditional intensity which is used in the covariance density is formed from the sum of the densities of the S_k. Note, however, that the log likelihood function for a set of consecutive intervals has a straightforward form and this is useful in fitting such a model to empirical data.

(v) *Autoregressive and moving average processes*

An important class of point processes specified via a dependent sequence of intervals involves autoregressive and/or moving average constructions. In each case the sequence of intervals is stationary with an exponential marginal distribution; such processes may prove useful simple alternatives to the Poisson process. Assume that $\{\varepsilon_i ; i = 0, 1, \ldots\}$ is a sequence of independent exponential variables with parameter ρ, and that $\{U_i ; i = 1, 2, \ldots\}$ and $\{V_i ; i = 1, 2, \ldots\}$ are independent of each other and of $\{\varepsilon_i\}$, and are sequences of independent and identically distributed Bernoulli variables with parameters $\alpha = \mathrm{pr}(U_i = 0)$ and $\beta = \mathrm{pr}(V_i = 0)$ respectively. Then we can define the following interval sequences

(i) $X_0 = \varepsilon_0$,

     $X_n = \alpha X_{n-1} + U_n \varepsilon_n \quad (n = 1, 2, \ldots)$;

(ii) $Y_n = \beta \varepsilon_n + V_n \varepsilon_{n-1} \quad (n = 1, 2, \ldots)$;

(iii) $Z_n = \beta \varepsilon_n + V_n X_{n-1} \quad (n = 1, 2, \ldots)$,

where X_n is given in (i).

The point process with interval sequence $\{X_n\}$ is called an exponential autoregressive process of order 1 (EAR 1), that with interval sequence $\{Y_n\}$ is an exponential moving average process of order 1 (EMA1), while that for which $\{Z_n\}$ forms the interval sequence is an exponential autoregressive-moving average process each of order 1, denoted EARMA (1,1). Because (i) and (ii) are special cases of (iii), we consider briefly the properties of (iii). Note that if $\alpha = \beta = 0$ the process is a Poisson process. It is possible to define higher order processes of each type but these will not be considered here.

The proof that the Z_n are marginally exponentially distributed is straightforward and is as follows. First,

$$E(e^{-sU_n \varepsilon_n}) = \alpha + (1 - \alpha)\rho/(\rho + s) = (\rho + s\alpha)/(\rho + s),$$

so that

$$E(e^{-sX_n}) = (\rho + s\alpha)(\rho + s)^{-1} E(e^{-s\alpha X_{n-1}}) \quad (n = 1, 2, \ldots).$$

Now $X_0 = \varepsilon_0$ so that

$$E(e^{-sX_n}) = \left(\frac{\rho + s\alpha}{\rho + s}\right)\left(\frac{\rho + s\alpha^2}{\rho + s\alpha}\right)\ldots\left(\frac{\rho + s\alpha^n}{\rho + s\alpha^{n-1}}\right) E(e^{-s\alpha^n X_0})$$

$$= \rho/(\rho + s).$$

Thus

$$E(e^{-sZ_n}) = E(e^{-s\beta\varepsilon_n - sV_n X_{n-1}})$$
$$= \rho/(\rho + s), \tag{3.22}$$

which shows that the intervals in the EARMA $(1,1)$ process are exponential with parameter ρ.

The autocovariances for the interval sequence are

$$c_k = \text{cov}(Z_n, Z_{n+k}) = \{(1-\beta)\alpha + \beta(1-\alpha)\}(1-\beta)\alpha^{k-1}/\rho^2$$
$$(k = 1, 2, \ldots); \tag{3.23}$$

see Exercise 3.6. For $k = 0$

$$c_0 = \text{var}(Z_n) = 1/\rho^2. \tag{3.24}$$

If $\alpha = 0$, the autocovariances for the EMA 1 process are

$$c_k = 0 \quad (k \geq 2), \qquad c_1 = \beta(1-\beta)/\rho^2, \qquad c_0 = 1/\rho^2,$$

while if $\beta = 0$, the autocovariance of lag k for the EAR 1 process is α^k/ρ^2 $(k \geq 0)$. The geometric form of the autocovariances is characteristic of linear Markov processes. Note that for the general EARMA $(1,1)$ process, $c_k \geq 0$ for all k, so that this class of models never gives negatively correlated intervals.

Higher order joint moments for the interval sequence can be obtained similarly. For the counting properties of a process which is constructed on the positive real line by taking the points to have coordinates $T_k = Z_1 + \ldots + Z_k$ $(k \geq 1)$, we need to know the distribution of T_k for all k, and this may be obtained as follows. Consider first the double Laplace transform $l_k(s_1, s_2)$ of the joint density of T_k and X_k, that is

$$l_k(s_1, s_2) = E(e^{-s_1 T_k - s_2 X_k}) \quad (k = 1, 2, \ldots). \tag{3.25}$$

Since $T_1 = Z_1$, it is straightforward to show, from the definitions of Z_1

and X_1, that

$$l_1(s_1, s_2) = \frac{\rho^2(\rho + s_1\beta + s_2\alpha)^2}{(\rho + s_1\beta)(\rho + s_2\alpha)(\rho + s_1\beta + s_2)(\rho + s_1 + s_2\alpha)}.$$

Also, for $k \geq 2$,

$$l_k(s_1, s_2) = E[\exp\{-s_1 T_{k-1} - (s_1 V_k + s_2\alpha)X_{k-1} - (s_1\beta + s_2 U_k)\varepsilon_k\}]. \tag{3.26}$$

Now ε_k is independent of T_{k-1} and X_{k-1}, and

$$E[\exp\{-(s_1\beta + s_2 U_k)\varepsilon_k\}] = \frac{\rho(\rho + s_1\beta + s_2\alpha)}{(\rho + s_1\beta)(\rho + s_1\beta + s_2)},$$

so that from (3.26)

$$l_k(s_1, s_2) = \frac{\rho(\rho + s_1\beta + s_2\alpha)}{(\rho + s_1\beta)(\rho + s_1\beta + s_2)}\{\beta l_{k-1}(s_1, s_2\alpha)$$
$$+ (1-\beta)l_{k-1}(s_1, s_1 + s_2\alpha)\}. \tag{3.27}$$

Thus, we can find the Laplace transform $l_k(s, 0)$ of the density of T_k by solving (3.27) recursively. If either $\alpha = 0$, for the EMA1 process, or $\beta = 0$, for the EAR1 process, the expressions for $l_k(s, 0)$ take a simple form. For example, if $\alpha = 0$,

$$l_k(s, 0) = \frac{\rho}{\rho + s}\left\{\frac{\rho(\rho + 2s\beta)}{(\rho + s\beta)(\rho + s + s\beta)}\right\}^{k-1} \quad (k = 1, 2, \ldots).$$

For the general EARMA $(1,1)$ model the expression for $l_k(s, 0)$ is much more complicated.

The second-order properties of counts of the process can be obtained from the conditional intensity $h(t)$ as described in Section 2.5, and since

$$h^*(s) = \int_0^\infty e^{-st} h(t)dt = \sum_{k=1}^\infty l_k(s, 0)$$

we can, in principle at least, derive these properties. For the EMA1 process it follows from the above form for $l_k(s, 0)$ that

$$h^*(s) = \frac{\rho(\rho + s\beta)(\rho + s + s\beta)}{s(\rho + s)(\rho + s\beta + s\beta^2)}$$

and therefore that

$$h(t) = \begin{cases} \rho + \dfrac{\beta(1-\beta)\rho}{(1-\beta-\beta^2)}\{e^{-\rho t} - e^{-\rho t/(\beta + \beta^2)}\} & (\beta + \beta^2 \neq 1), \\ \rho + (2\beta - 1)\rho^2 t e^{-\rho t} & (\beta + \beta^2 = 1). \end{cases} \tag{3.28}$$

Thus, at least for the EMA 1 process, the second-order properties are straightforward.

An important feature of the EARMA (1,1) process is that it is simple to simulate on a computer. Realizations of the process can, therefore, be generated, for example for use as input in a queueing or stock control model, or for comparison with empirical data.

(vi) *A process with independent locations*

The generalizations of the renewal process discussed above have all incorporated some dependence in the interval sequence. Another approach is as follows. The co-ordinate S_k of the kth point after the origin in an ordinary renewal process has distribution function $G^{(k)}$, the variables S_k ($k = 1, 2, \ldots$) being dependent. Suppose that a new point process is constructed by locating points independently so that the co-ordinate T_k of the kth point has distribution function $G^{(k)}$; thus points are no longer ordered in the sense that T_{k-1} need not be less than T_k, etc. Because of this, the process does not have simple interval properties. However, the distribution of the number of points $N(t)$ lying in $(0, t]$ can be studied. Let G have mean μ and variance σ^2. Then

$$N(t) = \sum_{k=1}^{\infty} \chi_k(t),$$

where

$$\chi_k(t) = \begin{cases} 1 & \text{if } T_k \in (0, t], \\ 0 & \text{otherwise}, \end{cases}$$

so that

$$E\{N(t)\} = \sum_{k=1}^{\infty} G^{(k)}(t)$$
$$\sim t/\mu$$

as $t \to \infty$, as shown in Section 2.5. Also

$$\text{var}\{N(t)\} = \sum_{k=1}^{\infty} G^{(k)}(t)\{1 - G^{(k)}(t)\}.$$

Since T_k has mean $k\mu$ and variance $k\sigma^2$, T_k will contribute appreciably to the sum if $G^{(k)}(t)$ is neither too close to zero nor to one, that is if

$$|t - k\mu| < c \sqrt{(k\sigma^2)}$$

for some fairly small c. Solving this inequality for k, we find that

$$|k - t/\mu| < c\sigma \sqrt{(t/\mu^3)} + O(1),$$

as $t \to \infty$, so that intuitively var $\{N(t)\} = O(\sqrt{t})$ as $t \to \infty$. In fact, weak conditions on G (Isham, 1975) are sufficient to ensure that as $t \to \infty$

$$\text{var} \{N(t)\} \sim \sigma \left(\frac{t}{\pi \mu^3} \right)^{1/2}.$$

Thus, for this process, the index of dispersion $I(t) \to 0$, whereas for a renewal process $I(t) \to \sigma^2/\mu^2$, as $t \to \infty$. Although $I(t) \to 0$ as $t \to \infty$, the process is not one of bounded variability (see Section 3.5), since var $\{N(t)\}$ is not bounded as $t \to \infty$.

The same conditions on G are also shown to be sufficient to ensure that, far from the origin, the process is approximately a Poisson process. This can be shown either directly or by using a limit theorem on the superposition of point processes which will be described in Section 4.5. Note that it is not necessary, in constructing this process, for G to be the distribution function of a positive random variable. Non-asymptotic results can also be obtained for the process. A particularly tractable special case is when G is the exponential distribution.

3.3 Simple intensity-based models

(i) *Introduction*
It has already been emphasized that one important way of specifying a point process, at least in one dimension, is via the complete intensity function or some generalization thereof. It is, therefore, natural to consider special models in which the intensity function takes a particularly simple form.

One type of process of importance in empirical statistical analysis is produced by introducing a dependence on an observed explanatory function or process. Thus suppose that at time t there is observed a vector $z(t)$ of explanatory variables. Then we can modify the simple intensity function $\rho(t; \mathscr{H}_t) = \rho$ of the Poisson process to, say,

$$\rho(t; \mathscr{H}_t) = \rho e^{\beta^T z(t)} \qquad (3.29)$$

for an externally modulated Poisson process. Here β is a vector of parameters with transpose β^T and of course modulating functions other than the exponential could be used. External dependencies can be introduced similarly into other processes specified by the complete intensity function; see also Section 1.3. Now so long as either $\{z(t)\}$ is non-random, or we condition on the observed $\{z(t)\}$, equation (3.29) defines a time-dependent Poisson process and no new probabilistic aspects arise. We shall, therefore, not study (3.29) in further detail here. If, however, the point process is controlled or influenced by a real-valued random process and the whole random system is observable and to be modelled, then we have a somewhat different situation to be considered in Chapter 5 under the general topic of marked point processes.

The orderly processes in Section 3.2 were specified simply in terms of intervals between successive points and have, as remarked previously, simple forms of complete intensity function. In particular, the complete intensity function for a renewal process depends only on the backward recurrence-time. In this section, in (ii) we consider processes in which $\rho(t; \mathcal{H}_t)$ depends linearly on the entire previous history of points. Then in (iii) we examine in more detail the doubly stochastic Poisson process introduced in Section 1.2(iv), in which the intensity depends on an unobserved stochastic process rather than, as in (3.29), on an observed non-random function.

(ii) Linear self-exciting processes

The linear self-exciting process is most concisely defined via the equation

$$\rho(t; \mathcal{H}_t) = \gamma + \int_{-\infty}^{t} w(t - z) dN(z), \tag{3.30}$$

where $\gamma > 0$ is a constant and w is a non-negative function such that

$$w^*(0) = \int_0^\infty w(x) dx < 1. \tag{3.31}$$

If we assume stationarity, it follows, on taking expectations in (3.30), that the rate ρ of the process satisfies

$$\rho\{1 - w^*(0)\} = \gamma. \tag{3.32}$$

We can regard (3.30) as specifying $E\{dN(t)|\mathcal{H}_t\}$. As such (3.30) has a formal connection with a linear autoregressive system and we can find an analogue of the Yule–Walker equations by multiplying (3.30) by $dN(t - \tau)$ for $\tau > 0$ and then taking expectations over $\mathcal{H}_t$. Thus assuming the stationarity of the process and using the formal result that

$$E\{dN(t)dN(t + x)\} = \rho\{h(x) + \delta(x)\} dt dx,$$

we have, for $\tau > 0$,

$$h(\tau) = \gamma + w(\tau) + \int_0^\infty w(z)h(\tau - z)dz.$$

Because h is an even function, this integral equation can be rewritten as

$$h(\tau) = \gamma + w(\tau) + \int_0^\tau w(z)h(\tau - z)dz + \int_0^\infty w(\tau + z)h(z)dz; \quad (3.33)$$

yet another form is in terms of the covariance density c. The most important special case has $w(z) = w_0 e^{-w_1 z}$, with $w_1 > w_0$ and, by (3.32), $\rho = w_1 \gamma/(w_1 - w_0)$.

If we take the Laplace transform of (3.33), we obtain, for this special case,

$$h^*(s) = \gamma/s + w_0/(w_1 + s) + w_0\{h^*(s) + h^*(w_1)\}/(w_1 + s). \quad (3.34)$$

The unknown constant $h^*(w_1)$ is determined by setting $s = w_1$. We then find, on solving (3.34) for $h^*(s)$ and inverting the transform, that

$$h(t) = \rho + \frac{w_0(2w_1 - w_0)}{2(w_1 - w_0)}e^{-(w_1 - w_0)t}. \quad (3.35)$$

A similar technique can be used when w is a linear combination of exponential terms; for a general solution (Hawkes, 1972) the more elaborate methods of Wiener–Hopf theory have to be used and the answer is most conveniently expressed in terms of the spectral density ψ of (2.37) associated with the second-order counting properties. In fact

$$\psi(\omega) = \frac{\rho}{2\pi|1 - \int_{-\infty}^\infty e^{ix\omega}w(x)dx|^2}.$$

In the special case when $w(z) = w_0 e^{-w_1 z}$, the defining intensity function, considered as a stochastic process, is a Markov process; its sample paths consist of deterministic exponential decays towards the asymptote γ interrupted by a state-dependent Poisson process of upwards jumps of amount w_0. Because of this Markov property, the associated point process may be called a Markov self-exciting point process.

In principle, various properties of the linear self-exciting process can be derived via the general results of Chapter 2. Unfortunately, simple explicit results do not emerge even for the Markov case. The variance of the number of points in an interval of length t can be deduced from the conditional intensity function of (3.35) via (2.27) and it is clear that the process is overdispersed relative to the Poisson process.

To obtain the distribution of intervals between successive points in the Markov case, let $\{R(t)\}$ be the stochastic process of values taken by the complete intensity function. Then if at $t = 0$, $R(0) = \rho_0$, the probability that there are no points in $(0, x]$ is

$$\exp\left[- \int_0^x \{\gamma + (\rho_0 - \gamma)e^{-w_1 z}\} \, dz \right]$$

$$= \exp\left\{ -\gamma x - \frac{(\rho_0 - \gamma)}{w_1}(1 - e^{-w_1 x}) \right\},$$

because of the exponential decay of the intensity function between points. To find the unconditional probability of zero points in $(0, x]$, we take expectations with respect to ρ_0 giving

$$\exp\left\{ -\gamma x + \frac{\gamma}{w_1}(1 - e^{-w_1 x}) \right\} f_R^*\left(\frac{1 - e^{-w_1 x}}{w_1} \right), \qquad (3.36)$$

where the Laplace transform or moment generating function refers to the equilibrium marginal distribution of $\{R(t)\}$. While this is, in principle, determined from the Kolmogorov equations for R, the answer is complicated. Numerical work via another approach (Oakes, 1975) shows that the intervals are overdispersed relative to the exponential distribution and that successive intervals have serial correlations that are fairly small and that decay only slowly.

The general difficulty of relating alternative types of specification of point processes has been mentioned a number of times. In the present case the defining equation (3.30) can be re-interpreted as a generalized Poisson cluster process; see Sections 3.1(iv) and 3.4. For, consider a Poisson process of cluster 'centres' of rate γ. With each centre associate a branching process of descendants as follows. First generation descendants form a time-dependent Poisson process of rate w; with each such descendant is associated a further similar process of second generation descendants, and so on. The whole process of descendants associated with a particular cluster centre is the superposition of all these subprocesses. The structure of this cluster process simplifies somewhat in the Markov case.

The linear self-exciting process can be generalized in various ways, for example by regarding (3.30) as the first term of the Volterra expansion

$$\rho(t; \mathcal{H}_t) = \gamma + \int_{-\infty}^t w_1(t - z) \, dN(z)$$

$$+ \int_{-\infty}^t dN(z_1) \int_{-\infty}^t dN(z_2) w_2(t - z_1, t - z_2) + \dots$$

or by introducing dependencies on observed real-valued processes (Brillinger, 1975).

(iii) *Doubly stochastic Poisson processes*

We now study a different generalization of the Poisson process in which there is an unobserved stochastic process $\{\Lambda(t)\}$ influencing the occurrence of points and such that if $\mathcal{T}^\Lambda$ denotes a whole realization of Λ, then

$$\rho(t; \mathcal{H}_t, \mathcal{T}^\Lambda) = \lambda(t), \tag{3.37}$$

where $\lambda(t)$ is the realized value of $\Lambda(t)$. That is, conditionally on $\mathcal{T}^\Lambda$ points occur in a time-dependent Poisson process.

In some ways it is physically more natural to introduce the process in terms of the history $\mathcal{H}_t^\Lambda$ of the rate process, as in Section 1.2(iv), rather than in terms of $\mathcal{T}^\Lambda$. If this is done, however, it is essential to add a requirement that Λ is not influenced by the occurrence of points, in the sense that for all t_0, the probabilistic properties of $\Lambda(t)$ for $t > t_0$, given $\mathcal{H}_{t_0}$ and $\mathcal{H}_{t_0}^\Lambda$, depend only on $\mathcal{H}_{t_0}^\Lambda$. We have chosen the specification (3.37) largely for simplicity of exposition.

An explanation of how such processes arise in applications has already been outlined in Section 1.2(iv). We call Λ the rate process.

Calculation of the properties of the doubly stochastic Poisson process usually proceeds via the corresponding property of the time-dependent Poisson process, found conditionally on $\mathcal{T}^\Lambda$, followed by passage to the unconditional form. Thus for any interval $(a, b]$

$$E\{N(a, b); \mathcal{T}^\Lambda\} = \int_a^b \lambda(u)\,du,$$

$$\text{var}\,\{N(a, b); \mathcal{T}^\Lambda\} = \int_a^b \lambda(u)\,du,$$

so that

$$E\{N(a, b)\} = \int_a^b E\{\Lambda(u)\}\,du, \tag{3.38}$$

and

$$\text{var}\,\{N(a, b)\} = \int_a^b E\{\Lambda(u)\}\,du + \text{var}\left\{\int_a^b \Lambda(u)\,du\right\}. \tag{3.39}$$

Therefore, the count is overdispersed unless $\int_a^b \Lambda(u)\,du$ is constant.

In the arbitrary interval $(0, x]$, the conditional probability of no points is

$$\exp\left\{ -\int_0^x \lambda(u)\, du \right\}$$

and, therefore, the unconditional probability can be obtained if the Laplace transform of the density of $\int_0^x \Lambda(u)\, du$ can be found. For a stationary process this leads in the usual way to the density of intervals between successive points.

In principle more detailed properties of the point process can be obtained from the probability generating functional. Conditionally on $\mathcal{T}^\Lambda$ this is given by (2.42) for a time-dependent Poisson process, with λ replacing ρ. Thus the unconditional probability generating functional is

$$E_\Lambda\left(\exp\left[-\int_{-\infty}^\infty \{1 - \xi(t)\} \Lambda(t)\, dt \right] \right). \tag{3.40}$$

It is natural to consider what can be determined about the structure of Λ given the structure of the point process. We shall see later in this subsection, that if Λ is stationary the conditional intensity of the point process determines the autocovariance function of Λ. Similar consideration of higher moment properties shows that if the joint distributions of the $\Lambda(t_i)$s are determined by their moments of all orders, then the process Λ is determined in principle from a specification of the point process. In fact, however, it can be shown that given the structure of a point process, there is at most one set of joint distributions for Λ such that the given point process is the corresponding doubly stochastic Poisson process (Kallenberg, 1976).

There is a considerable number of special cases that can be considered either because of mathematical simplicity or because of physical plausibility. Classification of such special processes is on the basis of the structure of Λ. First, and least interestingly, $\Lambda(t)$ may be a constant, λ say, within each realization. Then the process is specified by the univariate distribution determining the variation of rate from realization to realization. That is, each realization is a simple Poisson process, different realizations having different rates. Such a process is sometimes called a mixed Poisson process and its properties are easily derived. An immediate generalization arises if the time axis is divided into fixed and known disjoint sets $A_1, A_2, \ldots$, such that within A_i there is a Poisson process of rate λ_i, where $\lambda_1, \lambda_2, \ldots$ are values of independent and identically distributed random variables. The most natural case is where the sets $A_1, A_2, \ldots$ are adjacent intervals of the same length.

Next there are various forms of non-stationary process of which the most important are finite, i.e. contain, with probability one, only a finite number of points in, say, $(0, \infty)$. A special class of these is obtained from a Poisson process of fixed rate λ by 'termination' after an unobserved random time L, independent of the points in the originating Poisson process, and having density $f_L(l)$. Clearly other point processes can be converted into finite processes by this or more general termination mechanisms. Thus, under this particular model, Λ is the random function

$$\Lambda(t) = \begin{cases} \lambda & (0 \leq t \leq L), \\ 0 & (L < t). \end{cases}$$

As noted in Section 1.4(v) if, in particular, L has an exponential distribution with parameter θ, it is easily shown that the total number of points observed has a geometric distribution and most other properties of the process are easily calculated; see Exercise 1.11. The process is equivalent to one in which given that a point has occurred at time t, and given also the history $\mathcal{H}_t$, then the point at t is the last with probability $\theta/(\theta + \lambda)$.

The third and most important class of doubly stochastic Poisson processes derives from stationary processes Λ. Suppose to begin with that Λ is second-order stationary with mean μ_Λ and autocovariance function γ_Λ. It then follows directly from (3.38) and (3.39) that $E\{N(a, b)\}$ and $\text{var}\{N(a, b)\}$ depend only on $b - a$ and that, with $N(t) = N(0, t)$,

$$E\{N(t)\} = \mu_\Lambda t, \quad \text{var}\{N(t)\} = \mu_\Lambda t + 2\int_0^t (t - u)\gamma_\Lambda(u)\, du. \quad (3.41)$$

It follows by comparison with the general formula (2.27) that $\rho = \mu_\Lambda$ and

$$h(t) = \mu_\Lambda + \gamma_\Lambda(t)/\mu_\Lambda. \quad (3.42)$$

Because, given Λ, the number of points $N(t)$ has a Poisson distribution, the unconditional distribution of $N(t)$ can be found provided that the density, or its Laplace transform, for

$$\mathscr{I}\Lambda(t) = \int_0^t \Lambda(u)\, du$$

can be obtained in manageable form. Indeed the probability generating function of $N(t)$ is

$$G_{N(t)}(z) = f^*_{\mathscr{I}\Lambda(t)}(1 - z). \tag{3.43}$$

This result can be generalized to deal with the joint distribution of numbers in non-overlapping intervals.

To obtain the distribution of the intervals between successive points, we again argue first conditionally on Λ. It follows that

$$p_0(x) = \text{pr}\{N(0, x) = 0\} = E_\Lambda[\exp\{-\mathscr{I}\Lambda(x)\}], \tag{3.44}$$

so that, using (2.11), the density of the interval between successive points is

$$\frac{1}{\rho} \frac{d^2 p_0(x)}{dx^2}.$$

To particularize further, we need to specify the process Λ in more detail. We mention only a few of the many possibilities. Of these perhaps the simplest mathematically and the most plausible empirically, is to suppose that Λ is a Gaussian process with $\sigma_\Lambda/\mu_\Lambda \ll 1$, so that the chance of meaningless negative values for Λ can be ignored. Then $\mathscr{I}\Lambda(t)$ is normally distributed with mean $\mu_\Lambda t$ and variance

$$v_\Lambda(t) = 2 \int_0^t (t - u)\gamma_\Lambda(u) \, du.$$

Therefore by (3.43) the probability generating function of $N(t)$ is

$$\exp\{-\mu_\Lambda t(1 - z) + \tfrac{1}{2}(1 - z)^2 v_\Lambda(t)\}. \tag{3.45}$$

Now the modified Hermite polynomials $H_r^\dagger(x)$ have the generating function

$$\sum_{r=0}^\infty z^r H_r^\dagger(x)/r! = \exp(xz + \tfrac{1}{2}z^2).$$

It follows by comparison with (3.45) that

$$\text{pr}\{N(t) = r\} = \exp\{-\mu_\Lambda t + \tfrac{1}{2}v_\Lambda(t)\} \frac{\{v_\Lambda(t)\}^{r/2}}{r!} H_r^\dagger \left\{\frac{\mu_\Lambda t - v_\Lambda(t)}{\sqrt{v_\Lambda(t)}}\right\}. \tag{3.46}$$

Further, by (3.44),

$$p_0(x) = \exp\{-\mu_\Lambda x + \tfrac{1}{2}v_\Lambda(x)\},$$

so that the density of intervals between successive points is

$$\mu_\Lambda p_0(x) \left[\left\{ 1 - \frac{1}{\mu_\Lambda} \int_0^x \gamma_\Lambda(u)\,du \right\}^2 + \frac{1}{\mu_\Lambda^2} \gamma_\Lambda(x) \right].$$

In particular, the value of the density at $x = 0$, instead of being μ_Λ for an exponential distribution of mean μ_Λ^{-1}, is $\mu_\Lambda (1 + \sigma_\Lambda^2/\mu_\Lambda^2)$. For the Gaussian process to remain positive with high probability it is necessary that $\sigma_\Lambda/\mu_\Lambda$ is small, say that $\sigma_\Lambda/\mu_\Lambda < 0.3$. Thus the departure at the origin from the exponential density with the correct mean is slight. Numerical work with $\gamma_\Lambda(x) = \sigma_\Lambda^2 e^{-\kappa x}$ shows that the log survivor function is convex from below and that the departure from linearity is small.

We now mention briefly a number of other special cases that have been examined in the literature. One possibility uses a shot-noise process for Λ, that is

$$\Lambda(t) = \gamma + \alpha \int_{-\infty}^t e^{-\kappa(t-u)}d\Gamma(u), \qquad (3.47)$$

where Γ is an unobserved Poisson process. Lawrance (1972) gives the main properties of a rather more general form of this process.

Another special case is sometimes appropriate for representing the occurrence of errors or defects. Suppose that Λ takes just two values and alternates between these in accordance with some stochastic mechanism; the two values can be interpreted as corresponding to relatively error-prone and error-free states. A special case has the lower error rate zero. For the durations of the periods spent in the two states, the simplest model is the alternating Poisson process, i.e. these periods are independently and exponentially distributed with means μ_0 and μ_1, corresponding to the rates 0 and λ. Gaver (1963) introduced this type of process; see also Lawrance (1972) who gives some numerical results.

Finally, in addition to the special cases outlined above there are a number of applications of doubly stochastic Poisson processes in physics in which an appropriate stationary process Λ is indicated by physical theory. For example Λ may be the square of an Ornstein–Uhlenbeck process of zero mean, or the sum of squares of several such processes. An Ornstein–Uhlenbeck process is a stationary Gaussian process with exponential autocovariance function and is thus the continuous analogue of a Gaussian first-order autoregressive process. In this way a process Λ is obtained closely associated with a Gaussian process, but having the essential nonnegativity required of a rate of occurrence. Macchi (1979) has reviewed special doubly stochastic Poisson processes arising in physics; see also Snyder (1975, Chapter 6).

3.4 Cluster processes

Cluster processes, which were mentioned briefly in Section 3.1(iv) and 3.3(ii), have the following general structure. There is a point process of cluster centres and to each cluster centre is associated a random number of points forming a subsidiary process or cluster, these subsidiary points being distributed about the cluster centre in some specified way. The cluster process then consists of the super-position of all the separate clusters, points belonging to the same cluster not being identified as such. The cluster centres are not themselves included in the cluster process, although no generality is lost thereby since a subsidiary cluster may be defined always to have a point at its cluster centre.

The situation of most theoretical interest is where the overlapping of different clusters is non-negligible; of course if such overlapping can be ignored most properties of the process can be studied rather directly. To make much progress with theoretical discussion it seems to be essential to assume that the distributions of points in different clusters are independent of one another and of the configuration of cluster centres. Then, given the positions of the cluster centres, the whole point process is a superposition of independent processes and therefore its simpler properties can be calculated via those of sums of independent random variables.

To specify a full cluster process we need to give the process of cluster centres and the mechanism for generating individual clusters. Formal results can be obtained in some generality. For example, if N_c denotes counts connected with the process of cluster centres and if $N_{(t)}(A)$ is the number of subsidiary points in A arising from a cluster given to have centre at t, then for the total number of points $N(A)$ in A, we have

$$E\{N(A)\} = \int E\{N_{(t)}(A)\} E\{dN_c(t)\}. \tag{3.48}$$

This is a consequence of the additive property of expectations and holds provided only that the expected number of cluster points per centre does not depend on the multiplicity, $dN_c(t)$.

Given, however, the stronger assumptions we can generalize (3.48) first to the variance and higher cumulants of $N(A)$ and then to the probability generating functional of the point process. For this last, suppose that cluster centres have occurred at points $\{t_i\}$ $(i = 0, \pm 1, \ldots)$. Then, because of the independence of the separate clusters, the conditional probability generating functional of the point process is

$$\prod_{i=-\infty}^{\infty} E\left(\exp\left\{\int \log \xi(t)\, dN_{(t_i)}(t)\right\}\right)$$

$$= \prod_{i=-\infty}^{\infty} G_s[\xi; t_i], \qquad (3.49)$$

where $G_s[\xi; t_i]$ is the probability generating functional for a cluster centred at t_i. It follows on taking expectations that unconditionally

$$G[\xi] = G_c[G_s[\xi; .]], \qquad (3.50)$$

where G_c refers to the process of cluster centres, since from the definition (2.41)

$$G_c[\xi] = E\left\{\prod_i \xi(T_i)\right\}.$$

As with so many of these very general relations, (3.50) yields useful explicit results only for rather special situations. Most cluster processes that have been either studied in detail theoretically, or which have been proposed as plausible for particular applications, have the following further properties:

(a) they are stationary in that the process of cluster centres is stationary and the clusters relative to their cluster centres are identically distributed, and more particularly in that

(b) the process of cluster centres is a Poisson process.

A little progress can in fact be made by replacing (b) by, say, the assumption that the cluster centres form a renewal process, but we restrict ourselves here to processes satisfying both (a) and (b), i.e. to so-called stationary Poisson cluster processes.

To proceed we need, then, to specify the distribution of the number M of subsidiary points in a cluster; let its probability generating function be

$$G_M(z) = E(z^M) = \sum_{m=0}^{\infty} g_m z^m.$$

Incidentally note that if $M = 0$ the cluster is empty, so that a process with a rate ρ_c for the cluster centres and $g_0 > 0$ is equivalent to a process with the rate modified to $\rho_c(1 - g_0)$ and with all clusters non-empty. Having specified the distribution of M, we must also, of course, specify the positions of the points relative to the cluster centre.

The two main Poisson cluster processes studied in applications have already been introduced in Section 3.1(iv); particular applications may well, however, support other possibilities as appropriate. In the Neyman–Scott process the M points in a cluster are independently and identically distributed around the cluster centre with some probability density function f, whereas in the Bartlett–Lewis process the intervals between successive points in a cluster are independent and identically distributed with probability density function $\tilde{f}$, the clusters thus forming finite renewal processes, terminated by a mechanism operating in terms of numbers of points rather than in time. In either case there may also be a point at the cluster centre. Thus if clusters could be isolated it would usually be fairly easy to discriminate between a Neyman–Scott process with a non-uniform f and a Bartlett–Lewis process, provided that the clusters are of reasonable size, although this would not be quite so easy with a nearly uniform f.

The Neyman–Scott process, but not the Bartlett–Lewis process, generalizes immediately to more than one dimension. Both processes have been used in a number of applied fields. As a model of clusters of cars along a one-lane road neither model is really satisfactory, because in that application the natural clusters are not overlapping. A more convincing application of the Bartlett–Lewis process is in connection with computer failures, regarded as primary and secondary. Primary failures occur in a Poisson process and each leads to a finite renewal process of secondary failures. The secondary failures are recurrences of the same type as the primary failure, arising because of imperfect repair. Because distinct primary failures refer to different components in the computer, overlapping of distinct clusters can occur. Other applications of the processes are to ecological problems, in cosmology and to earthquakes; see, for example, Neyman and Scott (1958, 1972), Vere–Jones (1970) and Warren (1971).

The second-order counting properties of these Poisson cluster processes are derived via the conditional intensity function h. Two points located at x_1 and $x_2 > x_1$ must either

(i) belong to separate clusters, in which case they are placed independently; or

(ii) belong to the same cluster with centre at some t.

Thus for the Neyman–Scott process,

$$\text{pr}\,\{N(x_1, x_1 + \delta_1) = N(x_2, x_2 + \delta_2) = 1\}$$

$$= \{\rho_c E(M)\}^2 \delta_1 \delta_2 + \rho_c E\{M(M-1)\}$$

$$\times \int_{-\infty}^{\infty} f(x_1 - t)\, f(x_2 - t)\delta_1 \delta_2 dt + o(\delta_1 \delta_2),$$

since there are $M(M-1)$ ordered pairs of points in a cluster of size M which could have co-ordinates x_1 and x_2. Then the conditional intensity is

$$h(u) = \rho_c E(M) + \frac{E\{M(M-1)\}}{E(M)} \int_{-\infty}^{\infty} f(x)f(x+u)dx, \quad (3.51)$$

and the corresponding spectrum is

$$\psi(\omega) = \frac{\rho_c E(M)}{2\pi} + \frac{\rho_c E\{M(M-1)\}}{2\pi} |f^*(i\omega)|^2. \quad (3.52)$$

The variance–time function $V(t) = \text{var}\,\{N(t)\}$ can now be calculated via (2.27) and we find that

$$I(t) = \frac{V(t)}{E\{N(t)\}} \to \frac{E(M^2)}{E(M)} \geq 1, \quad . \quad\quad (3.53)$$

as $t \to \infty$, with equality holding if and only if all the clusters are of size one, in which case the process is a simple Poisson process. In fact (3:53) holds for all Poisson cluster processes. This can be seen informally by noting that, if t is large, then $N(t)$ is, except for end effects, the sum of all the cluster sizes for cluster centres in $(0, t]$, regardless of the positions of the points about their cluster centres. An extension of this argument can be used to prove the asymptotic normality of $N(t)$ as $t \to \infty$ for a broad class of cluster processes.

The results for the Bartlett–Lewis process analogous to (3.51) and (3.52) are

$$h(u) = \rho_c E(M) + \frac{1}{E(M)} E\left\{ \sum_{k=1}^{M} (M-k)\tilde{f}^{(k)}(u) \right\}, \quad (3.54)$$

where $\tilde{f}^{(k)}$ is the k-fold convolution of $\tilde{f}$, and

$$\psi(\omega) = \frac{\rho_c E(M)}{2\pi} + \frac{\rho_c}{2\pi} E\left[\sum_{k=1}^{M} (M-k)\{\tilde{f}^*(i\omega)\}^k \right]. \quad (3.55)$$

The spectrum (3.55) can be expressed in terms of the probability generating function G_M of M as

$$\psi(\omega) = \frac{\rho_c E(M)}{2\pi} + \frac{\rho_c}{2\pi} \left[\frac{\tilde{f}^*(i\omega)}{\{1 - \tilde{f}^*(i\omega)\}^2} \left[\{1 - \tilde{f}^*(i\omega)\} E(M) - 1 \right. \right.$$

$$\left. \left. + G_M\{\tilde{f}^*(i\omega)\} \right] \right]. \tag{3.56}$$

By comparison of (3.52) and (3.56) we see that the Neyman–Scott and Bartlett–Lewis processes are essentially different in character, for while the second order properties of the former involve only the first two moments of M, those of the latter use the complete distribution of M via its probability generating function.

The distribution of intervals between successive points can be obtained explicitly for both Neyman–Scott and Bartlett–Lewis processes, although the arguments are rather lengthy (Lawrance, 1972). The survivor function typically has an exponential tail and the distribution is overdispersed relative to the exponential distribution.

Other properties of these two processes, if required, are usually most simply determined by first conditioning on the process of cluster centres and then taking expectations. Alternatively we may write down the probability generating functionals for the processes and then derive particular properties from these as described in Section 2.7.

It follows from (3.50) and (2.42) that for each process the probability generating functional $G[\xi]$ satisfies

$$G[\xi] = \exp\left\{ -\rho_c \int_{-\infty}^{\infty} (1 - G_s[\xi; t]) \, dt \right\},$$

where ρ_c is the rate of the Poisson process of cluster centres. For the Neyman–Scott process the generating functional for a cluster with its centre at t may be written in closed form as

$$G_s[\xi; t] = E\left\{ \exp \int_{-\infty}^{\infty} \log \xi(u) \, dN_{(t)}(u) \right\}$$

$$= \sum_{m=0}^{\infty} g_m \int_{-\infty}^{\infty} du_1 f(u_1) \xi(t + u_1) \ldots \int_{-\infty}^{\infty} du_m f(u_m) \xi(t + u_m)$$

$$= G_M\left\{ \int_{-\infty}^{\infty} \xi(t + u) f(u) \, du \right\}. \tag{3.57}$$

For the Bartlett-Lewis process the generating functional for a cluster is

$$G_s[\xi; t] = \sum_{m=0}^{\infty} g_m \int_{-\infty}^{\infty} dv_1 \tilde{f}(v_1) \xi(t + v_1) \int_{-\infty}^{\infty} dv_2 \tilde{f}(v_2) \xi(t + v_1 + v_2)$$

$$\ldots \int_{-\infty}^{\infty} dv_m \tilde{f}(v_m) \xi(t + v_1 + \ldots + v_m),$$

but this cannot be written in closed form to give a simple expression for the generating functional $G[\xi]$ of the process, in spite of the straightforwardness of the construction.

Note that if the clusters are defined to have points at the cluster centres then the above expressions for $G_s[\xi; t]$ must be multiplied by an extra factor of $\xi(t)$.

The notion of clustering is admittedly vague and a number of other special processes can be thought of as cluster processes of a very special kind. For example the so-called unpunctuality process, to be studied in Section 3.5 as an example of a process of bounded variability, has points at $\{n + T_n; n = 0, \pm 1, \ldots\}$, where the Ts are independent and identically distributed displacements. Formally this amounts to a deterministic process of cluster centres, with clusters each having a single point. The qualitative properties of such a process are, however, so different from those of the processes studied above that the connection is best regarded as purely formal.

As has been remarked previously, equivalent special processes can be defined in superficially different ways. For example, the connection between the Markov linear self-exciting process and a cluster process has already been mentioned. There is a rather similar link between a special kind of Neyman–Scott process and a doubly stochastic Poisson process with a rate function Λ given by

$$\Lambda(t) = \int_{-\infty}^{t} w(t - u) \, dN_c(u), \tag{3.58}$$

where w is a suitably behaved weight function and N_c is a Poisson process. It is intuitively clear that this doubly stochastic Poisson process can be regarded as a Poisson cluster process in which the cluster centres are provided by the process N_c and clusters are independently and identically distributed about each centre. The cluster size has a Poisson distribution with mean $w^*(0) = \int_0^{\infty} w(u) du$ and the cluster points are independently distributed about their centre with probability density function $w(x)/w^*(0)$. Thus this process is a Neyman–Scott process with

$$G_M(z) = \exp\{-w^*(0)(1-z)\}$$

and

$$f(x) = w(x)/w^*(0).$$

3.5 Processes of bounded variability

Almost all the stationary processes studied so far are such that, for large t, $V(t) = \text{var}\{N(t)\} \sim bt$, for some positive constant b. The proportionality to t is a consequence of the virtual independence of the process over distinct long subperiods. Thus for any integer m

$$N(0, mt) = N(0, t) + N(t, 2t) + \ldots + N(mt - t, mt).$$

If for sufficiently large t the covariances of the random variables on the right-hand side can be ignored, it follows by stationarity that

$$V(mt) \sim mV(t),$$

which implies the asymptotic proportionality to t.

It seems clear that plausible models for specific applied problems will often have localized dependencies such that the above argument does indeed hold. Yet not all stationary point processes have $V(t) \sim bt$. A particular process for which $V(t) \sim O(\sqrt{t})$ was described in Section 3.2(vi), but possibly the most interesting exceptions arise when $V(t)$ is bounded; the process is then said to be of bounded variability.

Achievement of a bounded $V(t)$ demands some fairly strong imposed regularity in the process and we now sketch two rather different ways in which this might arise. In the first, the so-called unpunctuality process, points are displaced by independent amounts from $\{0, \pm a, \pm 2a, \ldots\}$, whereas the second is a state- and time-dependent Markov birth process.

In the unpunctuality process we have points 'timetabled' to occur in a regular sequence of spacing a, the actual time of occurrence for the point due to occur at na being $na + T_n$, where $\{T_n\}$ form a sequence of independent and identically distributed random variables with probability density function f. Such a process is strictly speaking not stationary, because of the special role of the origin; stationarity can be achieved by displacing the whole process by a single random amount Z, uniformly distributed over $(0, a]$, the effect of which is to randomize the phase of the process.

The unpunctuality process can be regarded as a degenerate form of cluster process in which the cluster centres are regularly spaced and in which each cluster has a single point. Either by the arguments outlined in Section 3.4, or from first principles, it follows that the probability generating functional for the process 'centred' on the origin is

$$\prod_{n=-\infty}^{\infty} \int_{-\infty}^{\infty} \xi(na+\tau)f(\tau)\,d\tau,$$

while that for the stationary or 'random phase' version is

$$a^{-1}\int_0^a dz \left\{ \prod_{n=-\infty}^{\infty} \int_{-\infty}^{\infty} d\tau\, \xi(na+z+\tau)f(\tau) \right\}.$$

To investigate the second-order properties, consider the conditional intensity $h(t)$, the rate of points at a time t after an instant at which a point is given to have occurred. We can without loss of generality take the given point as corresponding to $n = 0$, having actual arrival point τ_0, say. For the following argument we take the random phase Z fixed at, say, zero. Now given τ_0, the rate of points at $t + \tau_0$ is

$$\Sigma' f(t + \tau_0 - na),$$

where the sum is for all $n \neq 0$. Therefore

$$h(t) = \int_{-\infty}^{\infty} \left\{ \sum_{n=-\infty}^{\infty} f(t + \tau_0 - na) - f(t + \tau_0) \right\} f(\tau_0)\,d\tau_0. \quad (3.59)$$

Now the difference D between the displacements of two points has density

$$f_D(y) = \int_{-\infty}^{\infty} f(x)f(x+y)\,dx,$$

so that

$$h(t) = \sum_{n=-\infty}^{\infty} f_D(t - na) - f_D(t). \quad (3.60)$$

Finally, the sum in (3.60) is, by the Euler–McLaurin theorem, a close approximation to $a^{-1}\int_{-\infty}^{\infty} f_D(x)\,dx$ (Kendall, 1942; Moran, 1950), provided that f_D is smooth and has high contact with the axis at its extremes. Thus, with this approximation,

$$h(t) \simeq a^{-1} - f_D(t).$$

Therefore, by (2.27),

$$V(t) = \frac{t}{a} + \frac{2}{a} \int_0^t (t - u) h(u) \, du - \frac{t^2}{a^2}$$

$$\simeq \frac{t}{a} - \frac{2}{a} \int_0^t (t - u) f_D(u) \, du.$$

Now the distribution of D is symmetric about the origin so that, subject only to the existence of the first absolute moment of D,

$$V(t) \simeq \frac{2}{a} \int_0^\infty u f_D(u) \, du, \tag{3.61}$$

as $t \to \infty$, establishing the bounded variability property.

Other properties of the process can in principle be determined via the general relations of Chapter 2.

For our second special process with the property of bounded variability, we consider a specification in terms of the complete intensity function. At time t the random variable $L(t) = N(t) - \rho t$ is positive, zero or negative depending on whether the number of points occurring by time t is above, on or below a target ρt. We set $\rho = 1$ without loss of generality, by choosing a suitable time unit. In a Poisson process of unit rate, the complete intensity is always one. Suppose now that given $\mathcal{H}_t$ and in particular given that $L(t) = l$, the complete intensity is a function $g(l)$ of l only. We assume that g is, at least on the whole, a decreasing function of its argument such that $g(0) = 1$; note that the process $\{N(t)\}$ is Markovian. It is plausible that for suitable such functions g, there is an equilibrium distribution for $L(t)$ of zero mean and finite variance; the latter property implies that of bounded variability.

In fact a rigorous proof that $V(t)$ is bounded can be given for a rather general class of essentially decreasing g. Here we concentrate on finding the form of the equilibrium distribution of L, assumed to exist. The density $f(l; t)$ of the Markov process $L(t)$ satisfies the forward equation

$$\frac{\partial f(l; t)}{\partial t} = \frac{\partial f(l; t)}{\partial l} - g(l) f(l; t) + g(l - 1) f(l - 1; t), \tag{3.62}$$

since the sample path decreases at unit rate except for a unit jump upwards at each occurrence of a point.

If $L(0) = 0$, it is clear that it is possible for L to vanish exactly only at integer time points. There is thus a built-in periodicity, somewhat analogous to that in the unpunctuality process; it is most simply treated by supposing that $L(0)$ is given the stationary distribution $f(l)$ of the process, which must satisfy the differential-difference equation

$$0 = f'(l) - g(l)f(l) + g(l-1)f(l-1).$$

Explicit solution of this does not seem possible, even in special cases. Therefore we examine a diffusion approximation holding as the function g becomes very flat and hence the process L becomes very dispersed. For this it is useful to introduce a notional parameter ε and to consider a sequence of processes with associated functions

$$g_\varepsilon(l) = g(\varepsilon l),$$

where $g(0) = 1, g'(0) < 0$. Then the stationary density of $\Xi(t) = L(t)\sqrt{\varepsilon}$, which we denote by $q_\varepsilon(\xi) = f(\xi/\sqrt{\varepsilon})/\sqrt{\varepsilon}$, satisfies

$$\sqrt{\varepsilon}q_\varepsilon'(\xi) - g(\xi\sqrt{\varepsilon})q_\varepsilon(\xi) + g(\xi\sqrt{\varepsilon} - \varepsilon)q_\varepsilon(\xi - \sqrt{\varepsilon}) = 0. \qquad (3.63)$$

We now look for solutions of the form

$$q_\varepsilon(\xi) = q^{(0)}(\xi) + \sqrt{\varepsilon}q^{(1)}(\xi) + \varepsilon q^{(2)}(\xi) + \dots,$$

writing also $g(x) = 1 + g_1 x + \frac{1}{2}g_2 x^2 + \dots$.

The leading term of the expansion gives

$$\left\{ \frac{1}{2}(-g_1)^{-1}\frac{d^2}{d\xi^2} + \xi\frac{d}{d\xi} + 1 \right\}q^{(0)}(\xi) = 0,$$

yielding the solution $q^{(0)}(\xi) \propto \exp(g_1\xi^2)$, a normal density of variance $(-2g_1)^{-1}$. In the original notation we have shown that to this order of approximation $N(t)$ is normally distributed with mean t and variance $\{-2g'(0)\}^{-1}$. The approximation is likely to be good when the calculated variance is large compared with one. Probably the simplest special case is $g(l) = e^{-\kappa l}$, the variance being $(2\kappa)^{-1}$.

The above argument illustrates the application of diffusion approximations to point processes. It can be extended in various ways. For example, by taking further terms in the expansion, we may obtain more refined approximations to the equilibrium distribution and variance. To obtain the conditional intensity by a similar argument, examination of the transient solution is needed; see Exercise 3.11. The answer is that approximately

$$h(t) = 1 + \frac{1}{2}g'(0)\exp\{tg'(0)\}. \qquad (3.64)$$

3.6 Level crossings

Most of the special point processes studied above have been introduced directly via some plausible generating mechanism defined from first principles. Sometimes, however, point processes arise in connection with some other type of stochastic process and are then best studied via the theory of those stochastic processes.

A simple example is provided by an ergodic discrete-state time-homogeneous Markov process in continuous time, for example a generalized birth-death process with a stationary distribution. Choose a particular state as a reference state and suppose that a point is defined to occur whenever that state is entered. It is clear that the points form a renewal process; the calculation of the distribution of the time between successive points proceeds by defining a new Markov process in which the reference state is absorbing, the next time it is entered. Note, however, that the rate of the point process can be calculated directly as the ratio of the equilibrium probability of the state to the mean time spent in the state per visit, and these can usually be found simply. The essence of the matter is that each time the reference state is entered, the process 'starts afresh', independently of its past. That property might hold for a particular reference state without being true for all possible states; we would then have a non-Markov process with entry into the reference state forming a regeneration point, and the associated point process would still be a renewal process. Clearly interesting non-Markov processes could be used similarly to define special point processes that are not renewal processes.

Consider now the corresponding questions when the underlying process $\{Y(t)\}$ is real valued. To begin with suppose that $\{Y(t)\}$ has continuous 'well-behaved' sample paths. The most important special case is where $\{Y(t)\}$ is a stationary Gaussian process. If we take an arbitrary reference level y_0, we can define a point to occur when an upcrossing of level y_0 occurs, an upcrossing at t_0 being defined by the conditions $Y(t_0) = y_0$, $Y'(t_0) > 0$, for differentiable processes, Y' denoting the derivative with respect to time. Point processes can be defined somewhat similarly for downcrossings of the level y_0, for all crossings of level y_0 regardless of direction, and for the occurrence of a local maximum of the process regardless of the value of $Y(t)$.

Because the detailed study of point processes determined by level crossings leans heavily on the theory of the underlying process we shall not consider these point processes very fully here.

For a stationary Gaussian process we calculate the rate of the upcrossing point process as follows. We can without loss of generality take the process mean to be zero; write γ_Y for its autocovariance function, so that in particular $\gamma_Y(0) = \operatorname{var}\{Y(t)\}$. By writing $Y'(t)$ as the limit of $\{Y(t + \delta) - Y(t)\}/\delta$, it is easy to show that the joint distribution of any set of values of Y and Y' is multivariate normal and that

$$E\{Y'(t)\} = 0, \quad \text{cov}\{Y(t), Y'(t)\} = 0,$$
$$\text{var}\{Y'(t)\} = -\gamma_Y''(0), \quad \text{cov}\{Y(t+\tau), Y'(t)\} = -\gamma_Y'(\tau), \qquad (3.65)$$
$$\text{cov}\{Y'(t+\tau), Y'(t)\} = -\gamma_Y''(\tau).$$

In order for $Y(t)$ to be differentiable in mean square, i.e. for the limit of $\text{var}[\{Y(t+\delta) - Y(t)\}/\delta]$ to exist, we need γ_Y to be twice differentiable at the origin, implying $\gamma_Y'(0) = 0$, since γ_Y is an even function.

Now suppose that $Y(t) = y$, $Y'(t) = z$ and that we can treat the realization as effectively linear in $(t, t+\delta)$, i.e. that we have locally $Y(t+\delta_0) = y + z\delta_0$ for $0 < \delta_0 < \delta$. Then an upcrossing of y_0 occurs in $(t, t+\delta)$ if and only if $y_0 - z\delta < y < y_0$, with $z > 0$. Thus if $f_{Y,Y'}(y, z)$ denotes the equilibrium bivariate normal density of $\{Y(t), Y'(t)\}$, the rate of occurrence of upcrossings at y_0 is

$$\lim_{\delta \to 0+} \delta^{-1} \int_0^\infty dz \int_{y_0 - z\delta}^{y_0} dy\, f_{Y,Y'}(y, z) = \int_0^\infty z f_{Y,Y'}(y_0, z)\, dz, \quad (3.66)$$

and on inserting the bivariate normal form for the density, we have that the rate of upcrossings is

$$\rho_{y_0} = \frac{1}{2\pi} \left\{ -\frac{\gamma_Y''(0)}{\gamma_Y(0)} \right\}^{1/2} \exp\left\{ -\frac{y_0^2}{2\gamma_Y(0)} \right\}. \qquad (3.67)$$

Thus, in particular, the rate of upcrossings of the mean, zero, is proportional to the square root of minus the second derivative at the origin of the autocorrelation function of Y.

To try to obtain an upcrossing process that is a renewal process, it is, by the argument at the beginning of this section, sensible to take Y to be a Markov process, i.e. to consider the Ornstein–Uhlenbeck process with exponential autocorrelation function. Formal application of (3.67) gives an infinite rate. It can be shown that this reflects the strong local jitter of the Ornstein–Uhlenbeck process: when the process is near y_0 it crosses and recrosses the level indefinitely often in a very short time. Of course such properties are not physically realizable and the result should be interpreted as meaning that for physical processes closely described by the Ornstein–Uhlenbeck model, the rate of upcrossings is strongly influenced by the very short-term behaviour of the process as actually realized. It seems unlikely that there is any stationary Gaussian process and value of y_0 for which the process of upcrossings is exactly a renewal process; it can be shown, however, that as y_0 becomes large the upcrossing process tends to a Poisson process (Cramér and Leadbetter, 1967, p. 257).

In principle a combination of the type of argument sketched above with the general formulae of Chapter 2 yields further properties of the point process, for example the distribution of the interval between successive upcrossings. Unfortunately, the details tend to be complicated. Thus the conditional intensity function, $h(\tau; y_0)$, is calculated by applying (3.66) twice and using the four-variate normal distribution of $\{Y(t), Y'(t), Y(t+\tau), Y'(t+\tau)\}$, whose covariance matrix is determined by (3.65). We have that the probability of upcrossings of y_0 in both $(t, t+\delta_1)$ and $(t+\tau, t+\tau+\delta_2)$ is asymptotically

$$\delta_1 \delta_2 \int_0^\infty dz_1 z_1 \int_0^\infty dz_2 z_2 f(y_0, z_1, y_0, z_2), \qquad (3.68)$$

where for simplicity we have omitted the suffices on the joint density f. Therefore,

$$h(\tau; y_0) = \rho_{y_0}^{-1} \int_0^\infty dz_1 z_1 \int_0^\infty dz_2 z_2 f(y_0, z_1, y_0, z_2), \qquad (3.69)$$

where

$$f(y_0, z_1, y_0, z_2) = \frac{1}{(2\pi)^2 |\Omega|^{1/2}} \exp\left(-\tfrac{1}{2} y^T \Omega^{-1} y\right),$$

with $y^T = (y_0, z_1, y_0, z_2)$ and

$$\Omega = \begin{pmatrix} \gamma_Y(0) & 0 & \gamma_Y(\tau) & \gamma_Y'(\tau) \\ 0 & -\gamma_Y''(0) & -\gamma_Y'(\tau) & -\gamma_Y''(\tau) \\ \gamma_Y(\tau) & -\gamma_Y'(\tau) & \gamma_Y(0) & 0 \\ \gamma_Y'(\tau) & -\gamma_Y''(\tau) & 0 & -\gamma_Y''(0) \end{pmatrix}. \qquad (3.70)$$

Numerical integration from (3.69) and (3.70) is the basis of Fig. 3.2, which gives some qualitative idea of the upcrossing point process in the special case in which $\gamma_Y(t) = e^{-t^2/2}$. Note that as the level y_0 increases the function $h(\rho_{y_0}^{-1}\tau; y_0)$ rises relatively more rapidly to its limiting value, the rate of the process, illustrating the asymptotic Poisson character of the upcrossing process as y_0 increases. The smallness of the values of $h(\tau; y_0)$ for small τ means that it is very unlikely for two upcrossings to occur close together. This is a consequence of the very smooth form of the realizations of the process Y for this special autocorrelation function. From the conditional

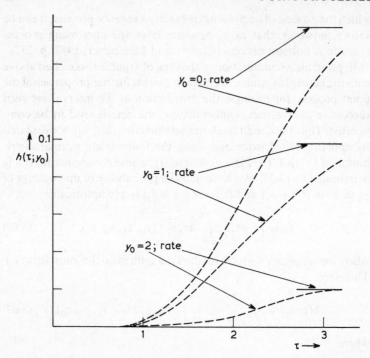

Fig. 3.2. *Upcrossings of level y_0 in stationary Gaussian process of zero mean, unit variance, autocorrelation function $e^{-\tau^2/2}$. ——, rate of process. ----, $h(\tau; y_0)$, conditional intensity function.*

intensity function other second-order properties, such as the variance of the number of upcrossings in $(0, t]$, are determined by the formulae of Chapter 2.

The results outlined above can be generalized in several directions and, of course, the proof under minimal regularity conditions of the existence of the main functions associated with very general crossing problems is mathematically interesting.

One other special case in which fairly simple explicit results are possible is obtained by taking for $Y(t)$ a generalized shot-noise process

$$Y(t) = \sum_{T_j < t} A_j e^{-\kappa(t - T_j)} ; \tag{3.71}$$

here T_j form a Poisson process of rate ρ_0 of impulses and $\{A_j\}$ are independent and identically distributed non-negative jump heights with density f_A. The process decays exponentially at rate $\kappa > 0$ between impulses. Upcrossings of a level y_0 can occur only at an impulse point and the rate of the upcrossing point process is

$$\rho_0 \int_0^{y_0} f_Y(y) \mathscr{F}_A(y_0 - y) \, dy, \tag{3.72}$$

where f_Y is the equilibrium density of Y and $\mathscr{F}_A$ is the survivor function for A. The rate of downcrossings is $\kappa f_Y(y_0)$, by essentially the same argument as used previously for Gaussian Y; the slope of the trajectories near y_0 is $-\kappa y_0$. It follows from the additive form of (3.71) that the equilibrium density of Y is determined by the moment generating function

$$E(e^{-sY}) = f_Y^*(s) = \exp\left[\rho_0 \int_0^\infty \{f_A^*(se^{-\kappa t}) - 1\} \, dt\right]. \tag{3.73}$$

Because of the relatively simple transient structure of the process Y, further properties of the upcrossing process can be determined. Thus to obtain the conditional intensity, we have, following the general argument used previously for Gaussian processes, that the probability of upcrossings of y_0 in $(t, t + \delta_1)$ and $(t + \tau, t + \tau + \delta_2)$ is asymptotically

$$\delta_1 \delta_2 \int_0^{y_0} dy \int_{y_0}^\infty du \int_0^{y_0} dz \rho_0 f_Y(y) f_A(u - y) f_Y(z; u, \tau) \rho_0 \mathscr{F}_A(y_0 - z), \tag{3.74}$$

where $f_Y(z; u, t)$ is the density of $Y(t)$ given that $Y(0) = u$. The required conditional intensity then follows from (3.72) and (3.74). The calculation of the transient behaviour of the shot-noise process (3.71) can be made first conditionally on the number M of impulses in $(0, \tau)$. If $M = 0$, $Y(\tau) = ue^{-\kappa \tau}$. If $M = 1$, we have one impulse uniformly distributed over $(0, \tau)$ and then subject to exponential decay; this gives a conditional Laplace transform of

$$\frac{1}{\tau} \int_0^\tau f_A^*(se^{-\kappa v}) \, dv.$$

It follows that

$$f_Y^*(s; u, \tau) = \exp(-sue^{-\kappa \tau}) \sum_{m=0}^\infty \frac{e^{-\rho_0 \tau}(\rho_0 \tau)^m}{m!} \frac{1}{\tau^m} \left\{ \int_0^\tau f_A^*(se^{-\kappa v}) \, dv \right\}^m$$

$$= \exp\left[-sue^{-\kappa \tau} + \rho_0 \int_0^\tau \{f_A^*(se^{-\kappa v}) - 1\} \, dv\right],$$

and the conditional intensity of the point process of upcrossings is in principle determined.

The results greatly simplify when the distribution of A is exponential; Weiss (1973) has given an elegant discussion of this case.

3.7 Concluding remarks

In this chapter we have outlined some special processes useful in applications and have illustrated the derivation of their simpler properties. The development of a model for a particular application raises the usual problems of model building, in particular the need for judicious simplification while retaining the essential elements of the phenomenon it is desired to represent. Sometimes one of the models discussed above will be directly useful; in other cases the special models may be a basis for further development or adaptation.

In some applications a model is developed from general considerations and then relevant data are examined for consistency with the model. Detailed discussion of statistical methods is outside the scope of this book; formal statistical methods for the examination of point process models are in any case not well developed, primarily because of the difficulty for most of the models of obtaining a likelihood function in useful form. Mostly, the choice of statistical methods for examining the more complex models is restricted further by the difficulty of computing the theoretical properties of the process. The general approach is thus to find theoretically a number of important properties and then to use some to fit unknown parameters and others to assess adequacy of fit. In other cases it may be sensible, having fitted parameters, to simulate synthetic 'data' from the model and to compare these rather informally with the real data. Empirical discrimination between alternative models with somewhat similar realizations is likely to be very difficult without extensive high quality data.

Formal accounts often presuppose that a model has been developed from subject-matter considerations and that the model has then to be checked against data. In fact, however, there is a two-way relation between the statistical analysis of data and model choice and it is reasonable to expect some guidance on model selection from empirical analysis of data.

Partly with this in mind, and partly to suggest some way of absorbing what tends to be a lot of confusing detail about special models, we now outline some of the main aspects of data and models which need consideration. These are as follows.

    (i) Stationarity should be examined, in the first instance of the rate of the process, but also of any important structure detected in the process. Non-stationarity may take the form of a general long-term change, sudden discontinuous changes or periodic fluctuations connected, for instance, with time of day.

(ii) Non-orderliness will involve modelling multiple occurrences at the same time instant.

(iii) The variance of the numbers of points in long intervals indicates general over- or under-dispersion relative to the Poisson process.

(iv) The conditional intensity function, for small or moderate values of its argument, helps to indicate local effects.

(v) The detailed form of the distribution of the numbers of points in short or moderate intervals may occasionally supplement the variance, which is implicit in (iv).

(vi) The frequency distribution of the intervals between successive points will usually be informative.

(vii) Serial correlations of low lags for intervals between successive points assess local dependencies in a way different from (iv).

If the data do not depart fairly clearly from a Poisson process, detailed interpretation and development of special models will often be premature.

It will quite often be possible on general subject-matter grounds to choose between processes in which intervals between successive points play a central role in formulation and those, for instance the doubly stochastic Poisson processes, or cluster processes, in which intervals are not so important. Then use of the most relevant of the above aspects will often lead to a suitable model. In the following we assume stationarity and orderliness. Then, for example, if intervals between successive points are judged important, it will be natural to look for a renewal process, or extensions thereof such as a semi-Markov process or a Wold process. Points (vi) and (vii) should indicate an appropriate model and predictions of aspects (iii) and (iv) used to check the adequacy of the model. If, however, intervals between successive points are not of central importance, (iii) and possibly higher-order generalizations of (iv), are likely to be the key to short-term behaviour, (vi) and (vii) being used to assess fit.

If, however, it is desired to use empirical data to choose between the two broad categories of model, it is not so clear how to proceed. If (vii) suggests the data to be consistent with a renewal process, that will often be the simplest model. A complicated form of dependence between successive intervals is a strong indication that a specification not directly based on intervals should be sought.

Table 3.1. Summary of properties

Process (Section)	(iii) $V(t)$, asympt.†	(iv) $h(t)$	(v) Distr. of $N(t)$	(vi) Intervals	(vii) c_k, Serial corr. of intervals
(a) Poisson (3.1)	t (all t)	const.	Poisson	Exptl	0
(b) Renewal (3.2.i)	t	(3.11)	(3.6)	Arbt.	0
(c) Semi-Markov (3.2.ii)	t	(3.17)	Ud or Od	Arbt.	0
(d) Moran (3.2.iii)	t (all t)	const.	Poisson	Exptl	$0(k>1)$
(e) Markov intervals (3.2.iv)	t	(3.28)	(3.21)	(3.18)	
(f) EARMA 1,1 (3.2.v)	$\sqrt{t}$		Ud or Od	Exptl	≥ 0
(g) Independent locations (3.2.vi)	t		Ud	Asymp. exptl	
(h) Self exciting (Markov) (3.3.ii)	t	(3.35)	Od	Od	
(i) Doubly stochastic (3.3.iii)	t	(3.42)	Od (3.43)	(3.44)	small, slow decay
(j) Cluster (3.4)	t	(3.51), (3.54)	Asymp. Od	Od	
(k) Unpunctuality (3.5)	1	(3.60)	Ud*	Od	

* In general not asymptotically normal for large t.   † The asymptotic power of t is shown.

Od, overdispersed; Ud, underdispersed; most properties given require regularity conditions and limiting and degenerate cases are excluded.

Bibliographic notes, 3

There is a very extensive literature on the properties of special point processes. Cox and Lewis (1966, Chapter 7) gave an introductory account with some emphasis on statistical analysis and Lawrance (1972) reviewed work up to the early 1970s, concentrating on processes with simple interval properties, on doubly stochastic Poisson processes and on cluster processes.

Khintchine (1960) was among the first to give a careful discussion of the Poisson process; see the text for references to a few of the many papers on characterization of the Poisson processes. The special discrete distributions arising from the Poisson distribution by assigning a distribution to the mean have a long history going back at least to Greenwood and Yule's (1920) study of industrial accidents; see Johnson and Kotz (1969, Chapter 8) for a review of these distributions.

For references on renewal processes, see Bibliographic notes, 1. Semi-Markov processes were introduced by Lévy (1956) and by Smith (1955) and a detailed study of properties made by Çinlar (1969), Pyke (1961a,b) and Pyke and Schaufele (1964, 1966). Wold (1948, 1949) introduced processes with Markov dependent intervals. Lai (1978) gives further results. The autoregressive and moving average processes of Section 3.2(v) are developed in papers by Jacobs and Lewis (1977) and Lawrance and Lewis (1977). The process with independent locations is due to Isham (1975). Linear self-exciting processes were studied by Hawkes (1971a, b, 1972) and the connection with cluster processes was set out by Hawkes and Oakes (1974). Doubly stochastic Poisson processes arise naturally from the considerations that led much earlier to the compound Poisson distribution. Cox (1955) studied them in the context of textile physics. Bartlett (1955) gave the probability generating functional and Grandell (1976) has described both the general mathematical theory and also certain forecasting and estimation problems. The important book of Snyder (1975) describes many special such processes arising in physics and engineering and discusses forecasting and control. The fractional Gaussian noise process of Mandelbrot (1977) could be used to define a doubly stochastic Poisson process with very long term fluctuations in rate. Kingman (1964) determined the circumstances under which a doubly stochastic Poisson process is a renewal process.

Cluster processes have been extensively studied and again arose out of univariate studies of the distribution of counts in fixed sets. Neyman and Scott (1958) were concerned with cosmology, Bartlett

(1963) primarily, although not exclusively, with traffic problems, and Lewis (1964) with computer failures.

The unpunctuality process of Section 3.5 was introduced in connection with tanker arrivals by Govier and Lewis (1961, 1963), Lewis (1961) and Lewis and Govier (1964). They called such processes ones of controlled variability; we have preferred the term bounded variability. The other process of Section 3.5, the birth–death process with self-correcting rate, has been studied by Isham and Westcott (1979) and Cox and Isham (1978).

Level crossings of Gaussian processes were examined in two pioneer papers by Rice (1944, 1945). Level crossings are of direct interest in a number of physical and engineering applications. Cramér and Leadbetter (1967) study the more mathematical aspects; for the connection with point processes, see particularly Leadbetter (1972b). Weiss (1973) gave level crossing properties in the shot-noise process.

Another point process associated with a real-valued process is that of records, a record occurring whenever all previous values of the process are exceeded; see Gaver (1976) and Westcott (1977) for general discussion.

Further results and exercises, 3

3.1. Show that, for a non-stationary Poisson process with rate $\rho(t)$, given that there are n points in an interval I, these points are independently and identically distributed on I with density $\rho(t)/\int_I \rho(u)du$.

Similarly, if T_n is the time of the nth point after the origin, show that given $T_n = t$, the $n - 1$ points in $(0, t)$ are independently distributed with density $\rho(u)/\int_0^t \rho(v)dv$. [Section 3.1(i).]

3.2. For the modified renewal process of Section 3.2(i) in which the interval to the first point has density g_1, show that the renewal density $h_m(t)$ is such that $h_m^*(s) = g_1^*(s)\{1 - g^*(s)\}^{-1}$. Verify that if and only if $g_1(x)$ has the special form $\mathcal{G}(x)/\mu_X$ for an equilibrium renewal process, then the renewal density is constant and equal to $1/\mu_X$. [Section 3.2(i); Cox (1962, Section 4.4)]

3.3. Formulate and investigate a time-dependent renewal process in which the dependence on time is via 'real time' rather than via the serial number of the interval. [Section 3.2(i).]

3.4. A point process is constructed from mutually independent random variables as follows. Positive integer-valued random variables $\{R_1^{(i)}, R_2^{(i)}, \ldots\}$ have the distribution $\{\pi_r^{(i)}\}$ with mean $\lambda_i (i = 1, 2)$ and positive continuous random variables $\{X_1^{(i)}, X_2^{(i)}, \ldots\}$ have densities $g^{(i)}$ and means and variances μ_i, σ_i^2 $(i = 1, 2)$. That is, there are four sequences each of independent and identically distributed random variables. Intervals between successive points are defined by taking $R_1^{(1)}$ intervals from the sequence $X^{(1)}$, then $R_1^{(2)}$ intervals from the sequence $X^{(2)}$, then returning to the $X^{(1)}$ sequence for a further $R_2^{(1)}$ intervals, etc. Under what circumstances is this process (a) a semi-Markov process, (b) an alternating renewal process. Prove that in equilibrium the intervals between successive points have density

$$\lambda_1(\lambda_1 + \lambda_2)^{-1} g^{(1)}(x) + \lambda_2(\lambda_1 + \lambda_2)^{-1} g^{(2)}(x)$$

and that the rate of the process is $(\lambda_1 + \lambda_2)(\lambda_1 \mu_1 + \lambda_2 \mu_2)^{-1}$. Show that the serial correlations between successive intervals can be computed recursively from μ_i, σ_i^2 and the two distributions $\{\pi_r^{(i)}\}$. [Section 3.2(ii); Ekholm (1972).]

3.5.   Prove that the function

$$\pi^{-1}(uv)^{-1/2} e^{-(u+v)} + \varepsilon(u-v)(2uv - u - v + 1) e^{-2(u+v)}$$

defined for $(u, v) \geq 0$ is, for sufficiently small ε, non-negative and represents a bivariate density with Laplace transform

$$(1 + s_1)^{-1/2}(1 + s_2)^{-1/2} + \varepsilon s_1 s_2 (s_1 - s_2)(1 + \tfrac{1}{2}s_1)^{-3}(1 + \tfrac{1}{2}s_2)^{-3}.$$

Suppose that $(U_1, V_1), (U_2, V_2), \ldots$ are independent vectors with the above distribution. Prove from the moment generating function that the random variables $U_1 + U_2$ and $V_1 + V_2$ have exponential distributions of unit mean but are not independent. Define a point process with points at $U_1 + U_2, V_1 + V_2, U_3 + U_4, V_3 + V_4, \ldots$ and discuss the relation between it and a Poisson process of unit rate. [Section 3.2(iii); James (1952).]

3.6.   Show that the intervals in the EARMA $(1,1)$ process $\{Z_n\}$ have autocovariances satisfying

$$\text{cov}\{Z_n, Z_{n+k}\} = c_k = (1 - \beta)^2 \text{cov}(X_{n-1}, X_{n+k-1})$$
$$+ \beta(1 - \beta) \text{cov}(\varepsilon_n, X_{n+k-1}),$$

where

$$\text{cov}(X_{n-1}, X_{n+k-1}) = \alpha^k/\rho^2, \text{cov}(\varepsilon_n, X_{n+k-1}) = \alpha^{k-1}(1 - \alpha)/\rho^2$$

and hence that (3.23) is satisfied. [Section 3.2(v).]

3.7.   Use the conditional intensity given in (3.28) to obtain the index of dispersion $I(t)$ for the EMA 1 process. [Section 3.2(v).]

3.8. Show from (3.35) that the Markov self-exciting process is overdispersed relative to the Poisson process. [Section 3.3(ii).]

3.9. Suppose that the random rate $\Lambda(t)$ of a doubly stochastic Poisson process is independent of t. Show that the intervals in the process have autocorrelations which do not depend on the lag, i.e.

$$\mathrm{corr}(X_i, X_{i+k}) = \mathrm{var}\,(1/\Lambda)\,\{\mathrm{var}\,(1/\Lambda) + E(1/\Lambda^2)\}^{-1}.$$

[Section 3.3(iii).]

3.10. Show that by randomizing the phase of the nonstationary Poisson process of Exercise 1.4 a stationary doubly stochastic Poisson process is obtained. Find its conditional intensity function. [Section 3.3(iii).]

3.11. For the Markov birth process of Section 3.5, solve the forward equation for

$$M_\varepsilon(t, \xi) = E\left[\Xi\left\{\frac{t}{-\varepsilon g'(0)}\right\}\,\bigg|\,\Xi(0) = \xi\right]$$

in terms of an asymptotic expansion as $\varepsilon \to 0$. Hence deduce the conditional intensity for the process N given in (3.64). [Section 3.5; Cox and Isham (1978).]

3.12. A process may be said to have positive dependence if for any two disjoint sets A_1, A_2, $\mathrm{cov}\,\{N(A_1), N(A_2)\} \geq 0$. Prove that for a stationary process with positive dependence, $V(t_1 + t_2) \geq V(t_1) + V(t_2)$, i.e. that the variance function V is superadditive. Prove that for any integer m, $V(mt) \geq mV(t)$. What does this imply about the asymptotic behaviour of V as $t \to \infty$? [Section 3.5.]

CHAPTER 4

Operations on point processes

4.1 Preliminary remarks

In this chapter we consider various operations that can be applied to convert one or more point processes into a new point process. The operations are:

(i) a change of time scale to produce, in particular, a Poisson process;

(ii) thinning, in which some of the points in the original process are deleted;

(iii) translation of individual points;

(iv) superposition, in which a number of separate processes are merged.

In studying (ii)–(iv), quite strong independence assumptions are for the most part made. For example, the simplest form of thinning arises when each point is deleted with a certain constant probability, independently of all the other points. However, some particular cases of dependent deletions are discussed in Section 4.3 in connection with electronic counters. Similarly for the simplest form of translation, the displacements of different points are independent and identically distributed random variables, independent of the point process; likewise for superposition we normally suppose the component processes to be independent.

Thinning arises in a variety of circumstances in which points are lost to observation. Translation arises where the points have random velocities, when there is jitter in the recording apparatus, and in a queueing situation where the arrival instant is 'translated' into a departure instant. If the congestion in the system is negligible, e.g. if the number of servers is large, the translation of a point is then its service-time and the assumption of independently and identically distributed displacements is reasonable.

Superposition of independent processes arises in various problems in neurophysiology, in telephone traffic with the superposition of calls from numerous subscribers and in physics, for instance when the process of emissions from a radioactive source is a superposition of processes deriving from separate atoms. Two categories of results are discussed in connection with these operations. The first includes properties of the transformed process and questions of invariance, while the second covers limiting results.

4.2 Operational time

In Section 3.1(ii) we described how a non-homogeneous Poisson process of rate $\rho(t)$ can be reduced to a Poisson process with unit rate by a simple non-linear transformation of the time scale, in which the new time variable τ is defined by

$$\tau(t) = \int_0^t \rho(u)\,du.$$

More generally, for a point process with complete intensity function $\rho(t; \mathscr{H}_t)$, the transformation given by

$$\tau(t; \mathscr{H}_t) = \int_0^t \rho(u; \mathscr{H}_u)\,du \qquad (4.1)$$

results in a Poisson process of unit rate. We call τ operational time. Note especially that τ is a deterministic function of time if and only if the process is a Poisson process; otherwise $\rho(u; \mathscr{H}_u)$ involves $\mathscr{H}_u$ and, therefore, the transformation is random, being dependent on the particular realization of the process. This randomness restricts the usefulness of the transformation; see, however, Aalen and Hoem (1978).

4.3 Thinning

In this section we suppose that points of a process with counting measure N are deleted to produce a thinned process with counting measure N_d. The simplest method is, with probability $1 - p$, to delete each point of N independently of the others and independently of the point process, and we discuss this case first.

Assume that N is a stationary orderly point process with rate ρ and conditional intensity function h. It is straightforward to write down the properties of the thinned process N_d in terms of those of N. For example N_d has rate ρ_d given by

$$\rho_d = p\rho,$$

and conditional intensity h_d, where

$$h_d(u) = ph(u). \qquad (4.2)$$

Either by integrating (4.2) appropriately, as described in Section 2.5, or by writing

$$N_d(A) = \sum_{i=1}^{N(A)} \chi_i,$$

where

$$\chi_i = \begin{cases} 1 & \text{with probability } p, \\ 0 & \text{otherwise,} \end{cases}$$

it is easily shown that

$$\text{cov}\{N_d(A), N_d(B)\} = p^2 \text{cov}\{N(A), N(B)\} + p(1-p)\rho |A \cap B|,$$
(4.3)

for arbitrary sets A and B.

If N is a renewal process with interval distribution F, then it follows immediately that N_d also is a renewal process such that its interval distribution F_d is a random convolution of F. The number of terms in the convolution is one more than the number of successively deleted points. Thus

$$F_d(x) = E\{F^{(1+\Delta)}(x)\},$$
(4.4)

where $\text{pr}(\Delta = n) = p(1-p)^n$ $(n = 0, 1, 2, \ldots)$.

Suppose now that N is not only thinned but also rescaled by a factor of p, so that the rate is unchanged. Thus the process N_d consists of the set of points $\{pt_j : t_j \text{ is not deleted}\}$, where $\{t_j\}$ is a realization of N. Equation (4.2) becomes

$$h_d(u) = h(u/p),$$

and

$$\text{var}\{N_d(t)\} = p^2 \text{var}\{N(t/p)\} + (1-p)\rho t.$$

Also, if N is a renewal process, then (4.4) becomes

$$F_d(x) = E\{F^{(1+\Delta)}(x/p)\},$$

with Δ as before. Thus the corresponding relation between the Laplace transforms of interval densities is

$$f_d^*(s) = \sum_{n=1}^{\infty} p(1-p)^{n-1} \{f^*(sp)\}^n$$

$$= \frac{pf^*(sp)}{1 - (1-p)f^*(sp)}.$$
(4.5)

It follows at once from (4.5) that any renewal process which is invariant under this combined operation of thinning and rescaling must have an interval distribution satisfying

$$f^*(s) = \frac{pf^*(sp)}{1 - (1 - p)f^*(sp)}, \tag{4.6}$$

for all $s \geq 0$. Define $q(s)$ by $q(s) = \{f^*(s)\}^{-1} - 1$, so that $q(s)$ satisfies

$$q(sp) = pq(s).$$

Then, for some μ,

$$q(s) = \mu s,$$

and thus

$$f^*(s) = 1/(1 + \mu s),$$

so that μ is the mean of the interval distribution. Thus the Poisson process is the only renewal process invariant under a thinning-rescaling operation of this sort.

When N is stationary and orderly but not necessarily a renewal process, an important aspect is the behaviour of the thinned-rescaled process of N_d in the limit as $p \to 0$. Intuitively, if the points are deleted independently with a probability very close to one, then in most cases there will be little dependence between the retained points, indicating a Poisson process as limit. If the initial process N satisfies

$$\lim_{|I| \to \infty} E\left\{ \left| \frac{N(I)}{|I|} - \rho \right| \right\} = 0 \tag{4.7}$$

uniformly over all time intervals I, then it is straightforward to show that the thinned-rescaled process N_d converges in distribution to a Poisson process as $p \to 0$. Westcott (1976) gives a simple proof, more general limiting results for thinned processes being due to Kallenberg (1975).

For any specific small value of p, it is important to know how closely the properties of N_d are approximated by those of the Poisson process. For example, the index of dispersion $I_d(t)$ can be written as

$$I_d(t) = \frac{\text{var}\{N_d(t)\}}{E\{N_d(t)\}} = 1 + \frac{2p^2}{t} \int_0^{t/p} (t/p - u)\{h(u) - \rho\}\, du$$

$$= 1 + 2p \int_0^\infty \{h(u) - \rho\}\, du + o(p)$$

$$= 1 + p\{I(\infty) - 1\} + o(p),$$

for fixed t as $p \to 0$, if $h(u)$ converges to the limit ρ sufficiently quickly, where $I(\infty)$ is the limiting index of dispersion of $N(t)$ as $t \to \infty$. Thus, as $p \to 0$, the index of dispersion of the deleted process converges to 1 from above if $I(\infty) > 1$, and from below if $I(\infty) < 1$.

So far in this section we have assumed that points are deleted independently. For a deletion scheme based on a Markov chain, see Exercise 4.2. We now consider some particular examples of dependent deletions which arise in the theory of electronic counters. In, for example, the study of radioactive isotopes, particles emitted by a radioactive source are counted electronically; however, only a subsequence of the original sequence of particles is recorded because the counter becomes blocked after it records a particle. The counter remains blocked for some interval during which no further particles can be counted. There are two common kinds of blocking. In the first, for a type I counter, the counter blocks each time a particle is recorded and during an interval in which the counter is blocked any further particles arriving have no effect on the counter. For a type II counter, every particle to arrive blocks the counter for an interval from the instant of arrival, only particles arriving when the counter is unblocked being recorded. Other blocking mechanisms are possible but will not be considered here.

Fig. 4.1 shows the sequences of points recorded by type I and type II counters for a particular input process in the case when the blocked (dead) time caused by a single particle is a constant τ.

We assume that the input point process N is a renewal process with interval sequence $\{Z_i\}$, where the variables Z_i have a common probability density function f_Z. The dead times are represented by the sequence $\{Y_i\}$ of independent identically distributed variables with the density function g. Each Y_i is the dead time caused by a single

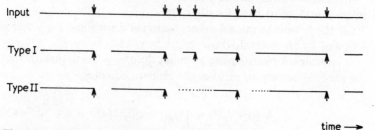

time $\longrightarrow$

Fig. 4.1. *Electronic counters.* ↓, *input process of points.*

Type I counter, ——*counter open.* ↑, *point recorded. Counter blocked following each recorded point.*

Type II counter, ——*counter open.* ↑, *point recorded.* ....., *blocked time extended by non-recorded point.*

arrival so that, for type II counters, several such dead times may contribute towards an interval during which the counter is continuously blocked. The output of the counter, that is, the recorded sequence, is the thinned process N_d and it is clear that for both type I and type II counters the deletions are dependent on the point process. Because of the structure of the system, the recorded process N_d is, in each case, a renewal process.

In practice, there may also be a completely random deletion of points because of counter inefficiency. However, since the effect of independent deletions on the input renewal process is to produce another renewal process, we shall assume that this effect has already been accounted for.

Simple special cases are when the input process N is a Poisson process and when the dead times Y_i are either constant or exponentially distributed. Because of the lack of memory implied in the behaviour of the counter, this last possibility is particularly tractable; however, it is physically unrealistic, because any electronic counter has a positive minimum dead time.

We consider first some properties of the process N_d for a type I counter. Denote the intervals in the renewal process of recorded events by $\{X_i\}$, where the X_i have probability density function f_X. Then f_X satisfies

$$f_X(x) = \int_0^x dy g(y) \left\{ f_Z(x) + \int_0^y dz h_Z(z) f_Z(x-z) \right\}, \qquad (4.8)$$

where h_Z is the renewal density corresponding to f_Z; see Section 3.2(i). For if $X = x$, then the interval X starts with a dead time of length y, for some $y < x$, and either

(i) no points of N are missed during $(0, y]$ in which case the first interval in N must have length x, or

(ii) the last of the missed points occurs at z, for some $z \le y$ and is followed by an interval whose length is $x - z$.

It is straightforward, using a similar decomposition, to show that the renewal density h_Z satisfies the renewal equation

$$h_Z(t) = f_Z(t) + \int_0^t f_Z(t-z) h_Z(z) dz. \qquad (4.9)$$

The use of (4.8) and (4.9) together enables us to write

$$F_X(x) = \int_0^x dy g(y) \int_y^x dz h_Z(z) \mathscr{F}_Z(x-z)$$

and, on integrating by parts, this simplifies to

$$F_X(x) = \int_0^x G(y)\mathscr{F}_Z(x-y)h_Z(y)dy, \qquad (4.10)$$

where F_X and $\mathscr{F}_X$ are the distribution function and survivor function of X, etc.

Since the output process N_d is a renewal process, the interval distribution (4.10) completely determines the process, and its properties can be obtained. However, for simple properties it may be easier to work directly. For example,

$$E(X \mid Y = y) = \mu_Z\{1 + H_Z(y)\},$$

since the mean number of complete intervals in $(0, y]$ in the process N is $H_Z(y)$, each with average length μ_Z, and we add 1 for the interval started before y and ending with the recorded point. Therefore,

$$\mu_X = E(X) = \mu_Z\left\{1 + \int_0^\infty H_Z(y)g(y)dy\right\}. \qquad (4.11)$$

Also if $p(t)$ is the probability that the counter is open at t given that it begins a dead time at $t = 0$, then $p(t)$ can easily be seen to satisfy

$$1 - p(t) = \mathscr{G}(t) + \int_0^t \mathscr{G}(t-u)h_X(u)du.$$

As $u \to \infty$, $H_X(u) \sim u/\mu_X$, so that as $t \to \infty$

$$p(t) \to 1 - \mu_Y/\mu_X.$$

Alternatively, since each X is the sum of a Y and an open period, the long term proportion of time during which the counter is blocked is μ_Y/μ_X.

In the special case when N is a Poisson process of rate ρ, results are particularly simple because $X = Y + V$, where V is a forward recurrence time in the Poisson process. Thus

$$E(X) = \mu_Y + 1/\rho,$$
$$\mathrm{var}(X) = \sigma_Y^2 + 1/\rho^2,$$

where σ_Y^2 is the variance of the distribution G, and therefore, using (3.9) and (3.10), $N_d(t)$ is asymptotically normally distributed with mean $\rho t(1 + \rho\mu_Y)^{-1}$ and variance $\rho t(1 + \rho^2\sigma_Y^2)(1 + \rho\mu_Y)^{-3}$. Hence, as $t \to \infty$,

$$I_d(t) = \frac{\mathrm{var}\{N_d(t)\}}{E\{N_d(t)\}} \to \frac{1 + \rho^2\sigma_Y^2}{(1 + \rho\mu_Y)^2}. \qquad (4.12)$$

In particular, if $\sigma_Y < \mu_Y$, which would almost always be the case in practice, then the limit in (4.12) is less than unity, showing that output process is, in the limit, underdispersed relative to the Poisson process.

We now consider some properties of the recorded process N_d for a type II counter. Because this situation is more complicated than that for the type I counter, we assume that the input process is a Poisson process. The counter will be open at t, given that it begins a blocked interval Y_0 at $t = 0$, if

(i) $Y_0 < t$ and

(ii) exactly n points are input in $(0, t]$ at $t_1, \ldots, t_n$ with corresponding dead times Y_i such that $Y_i < t - t_i$ $(i = 1, \ldots, n)$, for some $0 < t_1 < \ldots < t_n < t$ and some $n = 0, 1, \ldots$.

Thus the probability $p(t)$ that the counter is open at t satisfies

$$p(t) = G(t) \sum_{n=0}^{\infty} \int_0^t dt_1 \int_{t_1}^t dt_2 \ldots \int_{t_{n-1}}^t dt_n \rho^n e^{-\rho t} G(t - t_1) \ldots G(t - t_n)$$

$$= G(t) \exp\left\{ -\rho \int_0^t \mathcal{G}(t - u)\, du \right\}. \tag{4.13}$$

Then since the input is a Poisson process, the renewal density for the process N_d of recorded points is $h_d(t) = \rho p(t)$. Hence the interval density for N_d can be recovered by use of (3.8) which implies that

$$f_X^*(s) = \frac{h_d^*(s)}{1 + h_d^*(s)}.$$

Perhaps the simplest way to find the mean of the interval distribution is to note that $h_d(t) \to 1/\mu_X$ as $t \to \infty$. Therefore

$$1/\mu_X = \rho \exp\left\{ -\rho \int_0^{\infty} \mathcal{G}(u)\, du \right\}$$

and thus

$$\mu_X = \rho^{-1} e^{\rho \mu_Y}.$$

If we drop the assumption of a Poisson process input and assume instead that the dead times are constant, then again it is straightforward to find the interval distribution for the recorded renewal process via its renewal function; some results are given in Exercise 4.3.

4.4 Translation

In this section the points of a process with counting measure N are subject to independent and identically distributed translations, where we denote by f the translation density function. It follows immediately from the results on cluster processes in Section 3.4 that the probability generating functional of the translated process $\tilde{N}$ is given by

$$\tilde{G}[\xi] = G\left[\int_{-\infty}^{\infty} \xi(. + u) f(u) du\right], \qquad (4.14)$$

where G is the generating functional for N. All the properties of the process may in principle be obtained from (4.14), but for simple first- and second-order results a direct approach is preferable. Since the initial ordering of the points is likely to be lost on translation it is not particularly fruitful to consider the interval properties of the translated process in terms of those of the initial process and we restrict attention to the counting properties; see, however, Exercise 4.4. As before, we assume that N is a stationary orderly process.

The rate of the process and its stationarity and orderliness are unchanged by translation. Therefore, in particular,

$$E\{\tilde{N}(A)\} = \rho |A|, \qquad (4.15)$$

and the conditional intensity of $\tilde{N}$ is given by

$$\tilde{h}(u) = \int_{-\infty}^{\infty} h(u - v) f_D(v) dv, \qquad (4.16)$$

where f_D is the density of the difference D between two independent translations each with density f. Second-order properties of $\tilde{N}$ can be derived from (4.15) and (4.16) by the methods described in Section 2.5. In particular, it follows from (2.37) that the spectrum $\tilde{\psi}(\omega)$ of $\tilde{N}$ satisfies

$$\tilde{\psi}(\omega) - \rho/(2\pi) = \frac{\rho}{2\pi} \int_{-\infty}^{\infty} dv f_D(v) \int_{-\infty}^{\infty} du e^{-i\omega u} \{h(u - v) - \rho\}$$

$$= \{\psi(\omega) - \rho/(2\pi)\} f_D^{\dagger}(\omega), \qquad (4.17)$$

where ψ is the spectrum of N and

$$f_D^{\dagger}(\omega) = \int_{-\infty}^{\infty} e^{-i\omega v} f_D(v) dv.$$

Note that $f_D^{\dagger}(\omega)$ is real, by the symmetry of f_D. Thus the difference between the spectrum of the translated process and that for the Poisson process is given by the corresponding difference for the original process reduced by a factor which is the characteristic function of D.

Since the Poisson process is invariant under independent translations of its points, the question arises whether this property characterizes the Poisson process. It can be shown (Goldman, 1967) that the only process invariant under all sets of independent translations of its points is the mixed Poisson process, that is, a doubly stochastic Poisson process for which the random rate function $\Lambda(t)$ is a random constant Λ independent of t.

The limiting behaviour of the translated process $\tilde{N}$ is important. Intuitively, $\tilde{N}$ can be approximated by a Poisson process if, in translating the points, the dependencies between points in the original process have been effectively removed. This will be achieved if the translation distribution is suitably dispersed. The same result is obtained if the operation of translation is iterated a large number of times for a suitable density f. Formally, let N be a process satisfying the condition given in (4.7). Consider a sequence of translation distribution functions F_s and denote the translated process corresponding to F_s by $\tilde{N}_s$. If

$$\lim_{s \to \infty} \sum_{n = -\infty}^{\infty} |F_s(an + a) + F_s(an - a) - 2F_s(an)| = 0$$

for all $a > 0$, then as $s \to \infty$, $\tilde{N}_s$ converges to a Poisson process with rate ρ; for a proof of this, see Stone (1968).

As a particular instance of this limiting result, we consider the spectrum of $\tilde{N}$ given in (4.17). To study how (4.17) depends on the dispersion of the displacement distribution, it is convenient to write $f_D(v) = \tilde{f}_D(v/\sigma)/\sigma$, where σ is a scale parameter. Then

$$f_D^\dagger(\omega) = \int_{-\infty}^{\infty} e^{-i\omega\sigma x} \tilde{f}_D(x) dx.$$

By the Riemann–Lebesgue lemma, $f_D^\dagger(\omega) \to 0$ as $\sigma \to \infty$ for $\omega \neq 0$, and thus $\tilde{\psi}(\omega) \to \rho/(2\pi)$, the spectrum of a Poisson process. The rate of approach to zero depends on the analytic behaviour of $\tilde{f}$. If, in particular, $\tilde{f}$ is a standardized normal density function, then $f_D^\dagger(\omega) = e^{-\omega^2\sigma^2/2}$, so that $\tilde{\psi}(\omega) - \rho/(2\pi)$ is very small as soon as $\omega\sigma \gg 1$.

4.5 Superposition

We now consider the superposition of a number of independent processes. We have already noted in Section 2.7 that the generating functional for a superposition of independent point processes is simply the product of the individual functionals. As usual, though, for

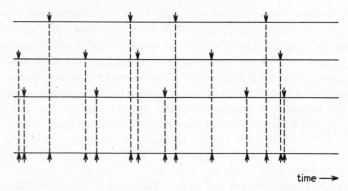

Fig. 4.2. *Superposition of three processes.* ↓, *points in component processes.* ↑, *points in pooled process.*

first- and second-order properties of the superposition it is simpler and more instructive to proceed directly. The qualitative effect of superposition is illustrated in Fig. 4.2; the superposed process is more like the Poisson process than the component processes.

Let $N_1, \ldots, N_k$ be the counting measures of k independent stationary, orderly processes where ρ_i, c_i and h_i denote, respectively, the rate, covariance density and conditional intensity of the ith process $(i = 1, \ldots, k)$. The superposition $N = N_1 + \ldots + N_k$ has rate ρ and covariance density c given by

$$\rho = \sum_{i=1}^{k} \rho_i,$$

$$c(u) = \sum_{i=1}^{k} c_i(u) = \sum_{i=1}^{k} \{\rho_i \delta(u) + \rho_i h_i(u) - \rho_i^2\}.$$

It then follows from (2.28) that the conditional intensity h satisfies

$$h(u) = \rho + \rho^{-1} \sum_{i=1}^{k} \rho_i \{h_i(u) - \rho_i\}. \tag{4.18}$$

Alternatively, (4.18) can be obtained by arguing that the probability density of having two points in the superposition a distance u apart is

$$\sum_{i=1}^{k} \rho_i h_i(u) + \sum_{i \neq j} \rho_i \rho_j, \tag{4.19}$$

where the first term is contributed by pairs of points belonging to the same process, while the second comes from pairs in which the two points come from distinct processes.

It follows from the way the superposition process is constructed that properties of counts of points will be more straightforward than those of intervals. However, it is easy to write down the marginal interval survivor function, $\mathscr{F}$, in terms of the interval survivor functions, $\mathscr{F}_i$, for the individual processes. The argument used in Section 1.2(ii) to find the density of the forward recurrence time in an equilibrium renewal process can be applied to any stationary orderly point process. It follows that the distribution of the forward recurrence time V_t in the superposition process satisfies

$$\text{pr}(V_t > x) = \prod_{i=1}^{k} \int_x^{\infty} \rho_i \mathscr{F}_i(v)dv, \qquad (4.20)$$

because V_t exceeds x if and only if the forward recurrence time in process i exceeds x, for all $i = 1, \ldots, k$. Thus, from (4.20), the interval survivor function $\mathscr{F}$ satisfies

$$\rho \mathscr{F}(x) = -\frac{d}{dx} \prod_{i=1}^{k} \int_x^{\infty} \rho_i \mathscr{F}_i(v)dv. \qquad (4.21)$$

It is clear that two identical processes cannot be superposed to obtain another process with exactly the same structure unless the superposition is combined with a rescaling operation similar to that used in studying deletions. For example, if two Poisson processes with the same rate are superposed and rescaled appropriately, the resulting process is another Poisson process with this rate.

Suppose, however, that we do not rescale the superposition and ask only that it should have the same general structure as the individual processes. Then, for example, as mentioned in Section 3.1(i), the superposition of two independent renewal processes is a renewal process if and only if they are Poisson processes; this applies to ordinary as well as to stationary renewal processes. Various generalizations (Çinlar, 1972) of this result have been to weaken the assumptions either about one of the original processes or about the superposition.

A simple proof in a very special case follows. Suppose that N_1 and N_2 are identically distributed independent ordinary renewal processes with interval survivor function $\mathscr{G}$ and corresponding density g such that g is continuous on $(0, \infty)$ and continuous on the right at the origin. Denote by f_{X_1, X_2} the joint density of the first two intervals X_1, X_2 in the superposition process N. Since either both of these intervals are intervals in the same process N_i $(i = 1, 2)$, or X_1 is an interval in N_i and $X_1 + X_2$ is an interval in N_j, $j \neq i$ $(i = 1, 2)$, then

$$f_{X_1, X_2}(x_1, x_2) = 2g(x_1)g(x_2)\mathscr{G}(x_1 + x_2)$$
$$+ 2g(x_1)g(x_1 + x_2)\mathscr{G}(x_2).$$

However, if N is also a renewal process, f_{X_1, X_2} must be symmetric in x_1 and x_2; that is,

$$g(x_1)g(x_1 + x_2)\mathscr{G}(x_2) = g(x_2)g(x_1 + x_2)\mathscr{G}(x_1),$$

and, letting $x_2 \to 0$, we find that

$$\{g(x_1)\}^2 = g(0)g(x_1)\mathscr{G}(x_1),$$

from which it follows that g is an exponential density.

There is an alternative approach which starts from the superposition N and then allocates a mark M_i to the ith point of N such that $M_i = j$ if the ith point of N belongs to the jth component process N_j, where here we assume that j takes values 1 and 2 only; more general marked point processes are discussed in Chapter 5. It is clear that if N is a Poisson process and the M_i are independent and identically distributed variables independent of N, then N_1 and N_2 are Poisson processes. Conversely, if the M_i are independent and identically distributed, independently of N, then N_1 and N_2 are independent only if N is a Poisson process; note here that N need not be stationary. For let $p = \mathrm{pr}(M_i = 1) = 1 - q$; then it follows from the independence of N_1 and N_2 that for any set A, the probability generating function G_N of $N(A)$ satisfies

$$G_N(z) = G_N(p + qz)G_N(q + pz).$$

Let $\tilde{G}(z) = d \log G_N(z)/dz$; then

$$\tilde{G}(z) = q\tilde{G}(p + qz) + p\tilde{G}(q + pz), \tag{4.22}$$

from which it follows that $\tilde{G}(z)$ is constant. To see this differentiate (4.22) n times with respect to z and evaluate it at $z = 1$. If $p \neq 0, 1$, i.e. neither N_1 nor N_2 is empty, then all the derivatives ($n \geq 1$) of $\tilde{G}(z)$ vanish at $z = 1$.

The asymptotic behaviour of the superposition of independent processes as the number of component processes tends to infinity has been extensively studied. Heuristically, we can argue as follows. Suppose that the points of each of the individual processes are suitably sparse, so that in any particular set A there will be, with high probability, at most one point from each process. Suppose also that no one process dominates the rest. Then $N(A)$ can be approximated by the sum of a large number of independent indicator variables and it follows that the distribution of $N(A)$ is close to a Poisson distribution. For example, if for each i, $\mathrm{pr}\{N_i(A) = 1\} = \rho_A/k$, then

$$E(z^{N(A)}) = \prod_{i=1}^{k} \left\{1 - \frac{\rho_A}{k}(1 - z)\right\} \tag{4.23}$$

and, as $k \to \infty$, the right-hand side of (4.23) approaches the generating function of a Poisson variable with parameter ρ_A. By a similar argument the dependence between counts in disjoint sets is negligible.

More formally, in an extremely clear exposition, Khintchine (1960) gave sufficient conditions for the superposition of independent stationary orderly processes to be asymptotically a Poisson process. A more general result (Grigelionis, 1963) which gives necessary and sufficient conditions for the superposition of arbitrary point processes to converge to a (not necessarily stationary) Poisson process is as follows. We consider a double array of point processes

$$N_{11}, \ldots, N_{1l_1}$$
$$\vdots$$
$$N_{k1}, \ldots, N_{kl_k}$$
$$\vdots$$

and let N_k be the superposition $N_{k1} + \ldots + N_{kl_k}$. We assume that $l_k \to \infty$ as $k \to \infty$, that for each k the processes N_{ki} $(i = 1, \ldots, l_k)$ are mutually independent, and that for any bounded interval I

$$\lim_{k \to \infty} \sup_{1 \le i \le l_k} \mathrm{pr}\{N_{ki}(I) \ge 1\} = 0.$$

Then, as $k \to \infty$, the superposition N_k converges to a Poisson process with mean measure P if and only if

$$\lim_{k \to \infty} \sum_{i=1}^{l_k} \mathrm{pr}\{N_{ki}(I) = 1\} = P(I),$$

and

$$\lim_{k \to \infty} \sum_{i=1}^{l_k} \mathrm{pr}\{N_{ki}(I) > 1\} = 0$$

for any finite interval I. The first of the above three conditions ensures that no processes are dominating the rest, while the third is concerned with the sparseness of the points of the individual processes. There are many similar limit theorems giving results for renewal processes and other special processes. One may also consider the convergence of superpositions of marked, multivariate or cluster processes to marked, multivariate or clustered Poisson processes.

We need to study not only the existence of a limiting process but also how rapidly that limit is approached. For example, suppose that the component processes in a superposition of stationary orderly processes are identically distributed. Suppose also that the intervals in the superposition are scaled by a factor of k so that their mean is the same as the mean interval in a component process. Then, from (4.21), the interval survivor function $\mathscr{F}$ in the superposition satisfies

$$\rho\mathscr{F}(x) = -\frac{d}{dx}\left\{\int_{x/k}^{\infty} \rho\mathscr{F}_i(v)dv\right\}^{k}$$

$$= \rho\mathscr{F}_i(x/k)\left\{1 - \int_{0}^{x/k} \rho\mathscr{F}_i(v)dv\right\}^{k-1}. \qquad (4.24)$$

Now if the common interval survivor function $\mathscr{F}_i$ in a component process has an expansion about the origin of the form

$$\mathscr{F}_i(x) = 1 - f_0 x - \tfrac{1}{2}f_1 x^2 + \ldots,$$

we can show, by substitution in (4.24), that

$$\mathscr{F}(x) = e^{-\rho x}\left\{1 + \frac{1}{k}(1 - f_0/\rho)(\rho x - \tfrac{1}{2}\rho^2 x^2) + O\left(\frac{1}{k^2}\right)\right\}.$$

An alternative approach is to suppose that the independent and identically distributed processes to be superimposed have conditional intensity h with an expansion $h(u) = h_0 + h_1 u + O(u^2)$ about the origin; let the mean interval be ρ^{-1} as before. Then in the scaled superposition the variance of the number of points in a period of length t, as $k \to \infty$, is

$$k\left[\frac{\rho t}{k} + 2\rho \int_{0}^{t/k} \left(\frac{t}{k} - u\right)\{h(u) - \rho\}\,du\right]$$

$$= \rho t + \frac{\rho t^2(h_0 - \rho)}{k} + \frac{\rho t^3 h_1}{3k^2} + O\left(\frac{1}{k^3}\right),$$

so that the index of dispersion is

$$1 + \frac{t}{k}(h_0 - \rho) + \frac{t^2 h_1}{3k^2} + O\left(\frac{1}{k^3}\right).$$

4.6 Infinite divisibility

In Section 4.5 we considered the superposition of k independent processes, relating properties of the superposed process to those of the individual processes. The number k was either fixed or, for the asymptotic results, very large. In this section, infinitely divisible processes are discussed; these are defined to be processes which, for each k, can be represented as a superposition of k independent and identically distributed processes. Thus it follows from Section 2.7 that the probability generating functional $G[\xi]$ of an infinitely divisible point process has, for each k, the form

$$G[\xi] = \{G_k[\xi]\}^k$$

for some generating functional $G_k[\xi]$. Clearly a Poisson process provides a simple example of such a process. For if N is Poisson with rate ρ then, for each k, N can be represented by $N = N_{k1} + \ldots + N_{kk}$, where the N_{ki} $(i = 1, \ldots, k)$ are independent Poisson processes each with rate ρ/k. The notion of infinite divisibility is relevant for model choice, because if the process of interest is formed by random superposition of a large number of component processes then an appropriate model should be infinitely divisible.

For a particular set A, if a process N is infinitely divisible, then the distribution of $N(A)$ must be an infinitely divisible distribution. It is well known (see for example, Feller, (1968, Section 12.2)) that an infinitely divisible discrete distribution is a compound Poisson distribution, that is, its probability generating function is of the form $\exp[-\mu\{1 - G(z)\}]$, where $G(z)$ is itself a probability generating function. Therefore by (1.21) $N(A)$ may be regarded as a random sum of independent variables with common probability generating function $G(z)$, where the number of terms in the sum is a Poisson variable with mean μ. This gives a direct physical interpretation of an infinitely divisible discrete distribution. Similarly, all the finite dimensional distributions of counts must be infinitely divisible and hence compound Poisson, with joint probability generating function of the form $\exp[-\mu\{1 - G(z)\}]$, where $G(z)$ is now a multivariate probability generating function.

Just as the property of infinite divisibility for the distribution of a random variable implies a particular form for the probability generating function, so a point process is infinitely divisible if and only if its generating functional has the form

$$G[\xi] = \exp\left[\int_{\mathcal{N}} \left\{\exp\int_{\mathbb{R}} \log \xi(t)\,d\tilde{N}(t) - 1\right\} d\tilde{P}(\tilde{N})\right], \quad (4.25)$$

any such representation being unique; see, for example, Matthes, Kerstan and Mecke (1978). In equation (4.25) $\mathcal{N}$ is the space of all counting measures $\tilde{N}$, and $\tilde{P}$ is a measure on $\mathcal{N}$ which satisfies two technical conditions, namely that it gives measure zero to an $\tilde{N}$ for which $\tilde{N}(\mathbb{R}) = 0$ and also $\tilde{P}\{\tilde{N}(I) > 0\}$ is finite for every finite interval I. Since $\tilde{P}$ need not be totally finite we cannot, in general, regard $\tilde{P}$ as a probability measure; however, if $\tilde{P}$ is totally finite with $\tilde{P}(\mathcal{N}) = \mu$, then we may write (4.25) in terms of the probability measure

$\tilde{Q} = \mu^{-1}\tilde{P}$ as

$$G[\xi] = \exp\{-\mu(1 - \tilde{G}[\xi])\},$$

with

$$\tilde{G}[\xi] = \int_{\mathcal{N}} \exp\left\{\int_{\mathbb{R}} \log \xi(t)\, d\tilde{N}(t)\right\} d\tilde{Q}(\tilde{N}).$$

Thus in this case, the infinitely divisible process N can be represented as a random superposition of independent processes, where the number of terms in the superposition has a Poisson distribution with mean μ.

An important special case arises if the measure $\tilde{P}$ is concentrated on finite processes $\tilde{N}$, i.e. if any $\tilde{N}$ with an infinite number of points is given zero weight. The infinitely divisible process is then said to be regular. In such a process the dependence between counts in disjoint sets can be made arbitrarily small if the sets are sufficiently widely separated, i.e. a form of mixing property holds. It can be shown that the class of all regular infinitely divisible point processes is the same as the class of all Poisson cluster processes (Section 3.4) with almost surely finite clusters.

An interesting class of point processes, introduced as a generalization of the Poisson process and defined through the probability generating functional, is the class of Gauss–Poisson processes. The generating functional of a Gauss–Poisson process has the form

$$G[\xi] = \exp\left[-\int_{\mathbb{R}} \{1 - \xi(t)\} P(dt) \right.$$
$$\left. + \frac{1}{2} \int\int_{\mathbb{R}^2} \{1 - \xi(t)\}\{1 - \xi(u)\} Q(dt \times du) \right], \quad (4.26)$$

where Q is, without loss of generality, a symmetric measure on $\mathbb{R}^2$, and where P and Q must satisfy suitable conditions in order that (4.26) represents the generating functional of a point process. If Q vanishes everywhere, then (4.26) reduces to the generating functional of a Poisson process. Note that P and Q are essentially just the first two moment measures for the process. For it is easily shown that

$$E\{N(A)\} = P(A),$$

and, if A and B are disjoint bounded sets, then

$$\text{cov}\{N(A), N(B)\} = Q(A \times B),$$

while all higher factorial cumulant measures are identically zero.

By considering separately the forms of (4.25) when $\tilde{P}$ gives its whole weight to realizations with either a single point or a pair of points, and comparing the results with (4.26), we see that the class of Gauss–Poisson processes is equivalent to the class of infinitely divisible processes for which the measure $\tilde{P}$ is concentrated on realizations which have either one or two points. Thus Gauss–Poisson processes are regular infinitely divisible processes and are, therefore, Poisson cluster processes.

Brillinger (1978b) gives the following representation of the Gauss–Poisson process which provides valuable insight into its behaviour as well as a way of simulating the process. Essentially, the Gauss–Poisson process can be regarded as the superposition of three one-dimensional processes, one of which is a Poisson process and the other two are the projections onto the co-ordinate axes of an independent Poisson process in the plane. More formally, let N_κ be a Poisson process in $\mathbb{R}^2$ with symmetric measure κ, that is

$$E\{N_\kappa(I \times J)\} = \kappa(I \times J),$$

for arbitrary intervals I and J, where $\kappa(I \times \mathbb{R})$ is finite. Thus the intensity of the planar process is such that the projection of the process onto either axis has a finite intensity.

Let N_L be a Poisson process on $\mathbb{R}$ with mean measure L. Define the process N^* on $\mathbb{R}$ to be the superposition of N_L and the projections of N_κ onto each of the axes. Then it is straightforward to show that the probability generating functional of N^* is

$$\exp\left[-\int_\mathbb{R} \{1 - \xi(t)\}\{L(dt) + 2\kappa(dt \times \mathbb{R})\} \right.$$
$$\left. + \int\int_{\mathbb{R}^2} \{1 - \xi(t)\}\{1 - \xi(u)\} \kappa(dt \times du) \right]. \qquad (4.27)$$

It follows by comparison of (4.27) with (4.26) that N^* is a Gauss-Poisson process. Conversely, a Gauss–Poisson process with generating functional given by (4.26) can be represented as a superposition in this way by taking

$$\kappa(I \times J) = \tfrac{1}{2}Q(I \times J),$$
$$L(I) = P(I) - Q(I \times \mathbb{R})$$

for all intervals I, J.

Bibliographic notes, 4

Aalen (1978) and Aalen and Hoem (1978) have described applications of operational time; for a general development, see Papangelou (1972) and Rudemo (1973).

Renyi (1956) proved some key results on the Poisson limit of thinned processes. Kallenberg (1975) discussed limiting results for thinned processes in a more general framework. There is an extensive literature on the theory of blocking in electronic counters; bibliographies are given by Smith (1958) and Müller (1975).

Translation of point processes was discussed by Cox (1963), invariance and limiting results were given by Goldman (1967) and more general limiting results by Stone (1968); dependent translations were considered by Harris (1971).

The earliest results on superposition are due to Palm (1943) and Khintchine (1960, Chapter 5), and an important limit theorem to Grigelionis (1963). For detailed study of some particular processes, see Cox and Smith (1953, 1954). Çinlar (1972) gives a comprehensive survey; see Lawrance (1973) for interval properties.

Infinite divisibility of point processes was studied by Kerstan and Matthes (1964), and Lee (1967) and discussed at length by Matthes, Kerstan and Mecke (1978). The fundamental measure $\tilde{P}$ of (4.25) for the general infinitely divisible point process is often called the KLM (Kerstan–Lee–Matthes) measure.

The Gauss–Poisson process was introduced by Newman (1970), and studied in detail by Milne and Westcott (1972) and by Brillinger (1978b).

Further results and exercises, 4

4.1. Show that the change to operational time, described in Section 4.2, results in a Poisson process of unit rate. [Section 4.2; Cox and Miller (1965) p. 153.]

4.2. Suppose that a stationary point process is thinned by a two-state Markov chain, i.e. suppose that the binary random variables χ_i of Section 4.3 are dependent, corresponding to the realization of a two-state Markov chain independent of the original point process. Find for large t the variance of the number of points in the thinned process in $(0, t]$ and comment on the effect of the dependence in the thinning process. Conjecture, and even prove, a theorem about a Poisson limiting process as the thinning probability tends to one; can non-Poisson limits be obtained? [Section 4.3; Isham (1980).]

4.3. Suppose that a type II counter with a constant dead-time τ and arbitrary renewal input, begins a blocked interval at $t = 0$. Show that, for $t > \tau$, $h < \tau$

$$E\{N_d(t, t+h)\} = Q(t+h) - Q(t),$$

where

$$Q(t) = F_Z(t) + \int_0^{t-\tau} \{F_Z(t-z) - F_Z(\tau)\} h_Z(z) dz.$$

Deduce that this result holds for all $h > 0$, and hence show that the renewal function for the process of recorded points satisfies

$$H_d(t) = E\{N_d(t)\} = H_Z(t) - \int_{t-\tau}^t F_Z(t-z) h_Z(z) dz$$

$$- F_Z(\tau)\{1 + H_Z(t-\tau)\}.$$

[Section 4.3.]

4.4.   A renewal process with interval density g and mean μ_X is subject to random translation in which with probability $1 - \varepsilon$ a point has zero displacement and with probability ε has a large displacement. Prove that for small ε the interval density in the translated process is, to order ε,

$$(1 - 2\varepsilon)g(x) - \varepsilon x g(x)/\mu_X + 2\varepsilon \mathscr{G}(x)/\mu_X + \varepsilon g^{(2)}(x),$$

where $g^{(2)}$ is the convolution of g with itself. Verify that the density is invariant for a Poisson process. Apply (4.17) to examine the effect on the spectrum. [Section 4.4.]

4.5.   Let $N_1, N_2, \ldots$ be independent and identically distributed equilibrium renewal processes. Define scaled superpositions $N^{(k)}$ for $k = 1, 2, \ldots$ by

$$N^{(k)}(t) = N_1(t/k) + \ldots + N_k(t/k).$$

By considering the probability generating function of $N^{(k)}$ in the limit as $k \to \infty$, show that, for large k, $N^{(k)}(t)$ has approximately a Poisson distribution. [Section 4.5.]

4.6.   Let N_1, N_2 and N_{12} be independent Poisson processes with rates ρ_1, ρ_2, ρ_{12}. The bivariate Poisson process consists of the processes $N_1 + N_{12}$ and $N_2 + N_{12}$. Find the joint probability generating functional of these processes and deduce that their superposition is a simple Gauss–Poisson process. [Section 4.6; Exercise 1.8; Milne (1974).]

CHAPTER 5

Multivariate point processes

5.1 Preliminary remarks

In most of the preceding discussion, we have considered processes in which the points are distinguished only by their instants of occurrence. Such processes may be called univariate point processes. We now turn to multivariate point processes in which two or more classes or types of point are observed. One example is of the arrivals and departures of customers in a queueing system and another is of the occurrences of two different kinds of electrical pulse in a neurophysiological problem.

Many of the ideas introduced in Chapter 2 generalize fairly directly; we review these in Section 5.2. The variety of special models is great and we discuss a few of the more interesting possibilities in Sections 5.3 and 5.4. One way of specifying a multivariate point process is to consider first the univariate point process of all points, regardless of class, and then to attach to each point a random variable M whose realized value gives the class, mark or type of the point in question. An obvious generalization is to allow M to be a general random variable. We then call the process a marked point process. Such processes are considered briefly in Section 5.5.

One important general distinction affects the choice of models both for particular situations and also, more especially, for statistical analysis. With two classes of point, it may be natural to regard say the class 2 points as dependent on, or even as caused by, the class 1 points. Then it will normally be desirable to consider the properties of the class 2 points conditionally on the positions of the class 1 points, i.e. to treat the process as univariate. The ideas studied in the present chapter, however, treat the classes of point symmetrically. The distinction is broadly that between regression and correlation in statistics.

The discussion in this chapter is deliberately focused on the description and characterization of processes rather than on the derivation of properties, which tends to be either fairly routine or extremely difficult.

5.2 Some general concepts

(i) *Definitions*

For much of the discussion we suppose for simplicity that there are just two classes of point. Let $N^{(1)}(A)$ and $N^{(2)}(A)$ be the numbers of points of the two classes in a set A and, in particular, write $N^{(1)}(t)$ and $N^{(2)}(t)$ when A is the interval $(0, t]$. We call $N^{(1)}$ and $N^{(2)}$ the marginal processes of class 1 and class 2 points and call N, the process of points taken without regard to class, the pooled or superposed process.

Throughout we consider only processes which are marginally orderly, i.e. ones for which both marginal processes are orderly in the sense defined in Section 2.3. If orderliness holds for the pooled process, we say that the process is orderly, or fully orderly. It is quite often useful in applications to consider marginally orderly processes but to allow simultaneous occurrences of class 1 and class 2 points. Such processes are, however, easily made fully orderly by defining such simultaneous occurrences to form a new class of points. It will usually be clear from the context whether we are dealing with orderly or only marginally orderly processes.

We restrict the discussion almost entirely to stationary processes. Complete stationarity is defined by direct extension of the definition of Section 2.2 requiring all probabilities of the form

$$\text{pr}\,\{N^{(1)}(A_1^{(1)}) = n_1^{(1)}, \ldots, N^{(1)}(A_j^{(1)}) = n_j^{(1)},$$
$$N^{(2)}(A_1^{(2)}) = n_1^{(2)}, \ldots, N^{(2)}(A_k^{(2)}) = n_k^{(2)}\}$$

to be invariant under arbitrary translations of the time axis. There is a considerable number of less stringent forms of stationarity which it would be tedious to list in detail. In essence, in studying any particular property of a point process we need only the stationarity of that and closely associated properties. Stationary orderly processes will be assumed to have finite rates ρ_1 and ρ_2 for the two classes of points: if the process is only marginally orderly we denote the rate of simultaneous class 1 and class 2 points by ρ_{12} and sometimes write $\rho_1 = \rho_{12} + \rho_1', \rho_2 = \rho_{12} + \rho_2'$.

Just as for univariate point processes, it is necessary to consider various choices of time origin for the process. The main possibilities are first that the origin is at an arbitrary time and secondly that the origin is taken at an arbitrary point of specified class. The latter possibility is sometimes called semisynchronous sampling. Where necessary we specify the origin by a subscript. Thus $N_a^{(b)}(t)$ is the number of class b points in $(0, t]$, given a class a point at the origin $(a, b = 1, 2)$; $N^{(b)}(t)$ is that number given a point of unspecified class at the origin.

Similarly, for $i = 1$, $X_a^{(b)}(i)$ denotes the time from the origin at a class a point to the next class b point and for $i = 2, 3, \ldots$ denotes the interval between successive subsequent class b points. If the subscript is omitted the time origin is at an arbitrary time so that, in particular, $X^{(1)}(1)$ and $X^{(2)}(1)$ form the pair of forward recurrence times from the origin to the first point of each class.

There are relations between counts and intervals corresponding to the univariate relations (1.7) and (2.2). For example, whatever conditions are imposed at the origin,

$$N^{(1)}(t_1) < n_1, \quad N^{(2)}(t_2) < n_2$$

if and only if

$$S^{(1)}(n_1) = X^{(1)}(1) + \ldots + X^{(1)}(n_1) > t_1,$$
$$S^{(2)}(n_2) = X^{(2)}(1) + \ldots + X^{(2)}(n_2) > t_2.$$

It is clear that a generalization of the argument used in Section 2.5 to prove asymptotic normality of counts in a univariate point process can now be employed to prove the asymptotic bivariate normality of $N^{(1)}(t_1)$ and $N^{(2)}(t_2)$ as t_1 and $t_2 \to \infty$, provided that the Xs satisfy weak conditions involving in particular the absence of long-term dependencies. The asymptotic covariance is expressed in terms of the cross-covariances of $X^{(1)}$ and $X^{(2)}$, assumed to be a stationary bivariate process.

(ii) *Specification of processes*

A recurring theme in dealing both with general theory and also with special processes is the interplay between different methods of specification. For univariate processes, these are via the complete intensity function, via counts and via intervals between successive points. For bivariate and, in general, multivariate point processes we have these three methods plus the fourth possibility of giving the univariate point process N, regardless of class, plus the structure of the class process, M.

For the complete intensity specification of an orderly process we consider the functions $\rho^{(1)}(t; \mathcal{H}_t^{(1)}, \mathcal{H}_t^{(2)})$ and $\rho^{(2)}(t; \mathcal{H}_t^{(1)}, \mathcal{H}_t^{(2)})$ defined in (1.22) and (1.23). If the process is only marginally orderly we need in addition the function $\rho^{(12)}(t; \mathcal{H}_t^{(1)}, \mathcal{H}_t^{(2)})$ for simultaneous occurrences of class 1 and class 2 points. It would then often be natural to modify the first two functions to $\rho^{(1\backslash2)}(t;\, \mathcal{H}_t^{(1)},\, \mathcal{H}_t^{(2)})$ and $\rho^{(2\backslash1)}(t;\, \mathcal{H}_t^{(1)},\, \mathcal{H}_t^{(2)})$ referring to occurrences of one class of point on its own. Thus

$$\rho^{(1)}(t;\, \mathcal{H}_t^{(1)},\, \mathcal{H}_t^{(2)}) = \rho^{(12)}(t;\, \mathcal{H}_t^{(1)},\, \mathcal{H}_t^{(2)}) + \rho^{(1\backslash2)}(t;\, \mathcal{H}_t^{(1)},\, \mathcal{H}_t^{(2)}).$$

(iii) *Conditional intensity functions*

The conditional intensity function h of (1.19) plays an important role in studying univariate point processes. An immediate generalization is to define, for $t \neq 0$,

$$h_b^{(a)}(t) = \lim_{\delta_1, \delta_2 \to 0+} \delta_2^{-1}\, \text{pr}\, \{N^{(a)}(t, t + \delta_2) > 0 \,|\, N^{(b)}(-\delta_1, 0) > 0\};$$

$$\tag{5.1}$$

$h_2^{(1)}$ and $h_1^{(2)}$ are called cross-intensity functions. In terms of the intervals introduced in (i), for orderly processes we have in an obvious notation that for $t > 0$

$$h_b^{(a)}(t) = \sum_{r=1}^{\infty} f_{S_b^{(a)}(r)}(t). \tag{5.2}$$

The definition (5.1) can be extended to $t = 0$ by including Dirac delta function components. The univariate conditional intensity functions $h_a^{(a)}$ are even functions of t and it follows from the multiplication law of probabilities that

$$\rho_b h_b^{(a)}(t) = \rho_a h_a^{(b)}(-t). \tag{5.3}$$

For processes without long term effects $h_b^{(a)}(t) \to \rho_a$ as $t \to \infty$.

It may sometimes be required to calculate the univariate conditional intensity function $h(t)$ for the superposed process in which the class of the points is ignored. Then for an orderly process, the point at the origin is class a with probability $\rho_a/(\rho_1 + \rho_2)$ and hence

$$h(t) = \frac{\rho_1}{\rho_1 + \rho_2} \{h_1^{(1)}(t) + h_1^{(2)}(t)\} + \frac{\rho_2}{\rho_1 + \rho_2} \{h_2^{(1)}(t) + h_2^{(2)}(t)\}. \tag{5.4}$$

Analogously to the univariate case, rather than using $h_b^{(a)}(t)$, for some purposes it is more convenient to use an equivalent function, the covariance density, defined for $t \neq 0$ by

$$c_b^{(a)}(t) = \lim_{\delta_1, \delta_2 \to 0+} (\delta_1 \delta_2)^{-1}\, \text{cov}\, \{N^{(a)}(t, t + \delta_2), N^{(b)}(-\delta_1, 0)\} \tag{5.5}$$

$$= \rho_b h_b^{(a)}(t) - \rho_a \rho_b, \tag{5.6}$$

with the usual extension of the definition when $t = 0$ for $a = b$. There is an associated spectral density

$$\psi_b^{(a)}(\omega) = \frac{1}{2\pi} \int_{-\infty}^{\infty} c_b^{(a)}(u) e^{-i\omega u} \, du. \tag{5.7}$$

One application of these functions for orderly processes is in the generalization of (2.27); we have that

$$C(t) = \text{cov} \, \{N^{(1)}(t), N^{(2)}(t)\}$$

$$= \text{cov} \left\{ \int_0^t dN^{(1)}(u), \int_0^t dN^{(2)}(v) \right\}$$

$$= \int_0^t (t - u) \{c_2^{(1)}(u) + c_1^{(2)}(u)\} \, du. \tag{5.8}$$

Under weak conditions, as $t \to \infty$,

$$C(t) \sim t \int_0^{\infty} \{c_2^{(1)}(u) + c_1^{(2)}(u)\} \, du$$

$$= t(\rho_1 \rho_2)^{1/2} I^{(1, 2)}, \tag{5.9}$$

where $I^{(1, 2)}$ is an asymptotic index of covariance generalizing the dispersion indices of (1.18) and (2.31). For large t, the correlation coefficient between $N^{(1)}(t)$ and $N^{(2)}(t)$ thus tends to a limit.

5.3 Some special processes

Having in the previous section outlined some general theory, we now turn to some special processes, describing some broad categories of process in the present section and discussing a very special application in Section 5.4. The variety of special processes is bewildering; for instance the majority of special processes in Chapter 3 have one or more multivariate forms. We discuss first in Section 5.3(i) processes with some fairly strong features of independence.

(i) *Notions of independence*

Most special models involve some independence relationships between the random variables associated with the process and it is helpful to begin by reviewing briefly some especially strong kinds of independence that may hold. We deal throughout with processes that are at least marginally orderly.

First there is the possibility that the class 1 and class 2 processes are completely independent in the strong sense that for all $t, \rho^{(1)}(t; \mathcal{H}_t^{(1)}, \mathcal{H}_t^{(2)})$ depends only on $\mathcal{H}_t^{(1)}$ and $\rho^{(2)}(t; \mathcal{H}_t^{(1)}, \mathcal{H}_t^{(2)})$ depends only on $\mathcal{H}_t^{(2)}$, the process also being orderly. The main properties of the superposed process can be determined, using for example some of the arguments deployed in Section 4.5 in connection with the superposition of independent processes.

Next there is the possibility that the identifying class process, $\{M\}$, is totally random, i.e. that the bivariate point process is in effect formed from some univariate point process of rate ρ by assigning the jth point of the process to class M_j, where M_j takes value a with probability ρ_a/ρ and $\rho_1 + \rho_2 = \rho$. Here the random variables $\{M_j; j = 0, \pm 1, \pm 2, \ldots\}$ are independent of one another and of the underlying univariate point process. Processes with completely random class is a reasonable general name for such processes. A particular consequence of the definition is that

$$h_b^{(a)}(t) = (\rho_a/\rho) h(t), \tag{5.10}$$

where $h(t)$ refers to the originating point process. Comparison of sample values with (5.10) would provide a simple empirical test for consistency with this family of processes. The family is extended in an obvious way to incorporate processes that are only marginally orderly.

The next special possibility is that one of the classes of points, say class 1, contains all the information concerning the future development of the process. We call the process purely class 1 dependent if, for all t, the complete intensity functions for all classes of point involve only $\mathcal{H}_t^{(1)}$. Obviously a variety of special processes can be constructed with this property. Just one example is provided by taking a Poisson process of class 1 points and then taking the complete intensity function for the class 2 points to be some given function $g(u_t^{(1)})$, say, of the backward recurrence time from t to the previous class 1 point. Some properties of this process are sketched in Exercise 5.3.

A final special kind of near-to-independence arises when the two classes of points occur independently, except for the possibility of simultaneous occurrences. That is $\rho^{(1 \backslash 2)}(t; \mathcal{H}_t^{(1)}, \mathcal{H}_t^{(2)})$ depends only on $\mathcal{H}_t^{(1)}$, $\rho^{(2 \backslash 1)}(t; \mathcal{H}_t^{(1)}, \mathcal{H}_t^{(2)})$ depends only on $\mathcal{H}_t^{(2)}$, but the intensity function for simultaneous occurrences may depend on both $\mathcal{H}_t^{(1)}$ and $\mathcal{H}_t^{(2)}$. The most natural special case arises when there are three mutually independent point processes, of single class 1 points, of single class 2 points and of simultaneous class 1 and class 2 points.

The simplest and most important illustration of several of these ideas is based on three independent Poisson processes of rates, say,

ρ_1', ρ_2', ρ_{12}, giving single class 1, single class 2 and combined class 1 and class 2 points, respectively. The process has completely random class; it is easily shown to be equivalent to a Poisson process of rate $\rho = \rho_1' + \rho_2' + \rho_{12}$ with completely random assignment of class with probabilities respectively ρ_1'/ρ, ρ_2'/ρ, ρ_{12}/ρ. The main properties of the process can be found fairly simply. Thus the numbers of points $(N^{(1)}(t), N^{(2)}(t))$ in $(0, t]$ have a bivariate Poisson distribution, by virtue of the definition of that distribution in terms of three independent random variables with Poisson distributions. The marginal processes are Poisson processes of rates $\rho_1 = \rho_1' + \rho_{12}$, and $\rho_2 = \rho_2' + \rho_{12}$. Finally, the forward recurrence times $V_1 = X^{(1)}(1)$ and $V_2 = X^{(2)}(1)$ from an arbitrary time to the first point of class 1 and of class 2, respectively, have the singular bivariate density

$$\rho_{12}e^{-\rho v_1}\delta(v_1 - v_2) + \begin{cases} \rho_1'\rho_2 e^{-\rho_1' v_1 - \rho_2 v_2} & (v_1 < v_2), \\ \rho_1\rho_2' e^{-\rho_1 v_1 - \rho_2' v_2} & (v_1 > v_2). \end{cases} \quad (5.11)$$

Milne (1974) showed that this process is a special case of the family of bivariate infinitely divisible processes with Poisson marginals.

(ii) *Doubly stochastic, cluster and linear self-exciting processes*
Doubly stochastic Poisson processes, cluster processes and linear self-exciting processes all have immediate bivariate generalizations. For the first, we have simply to replace the univariate rate function Λ by a vector $(\Lambda^{(1)}, \Lambda^{(2)})$ and, where it is required to have simultaneous occurrences, to include a third process $\Lambda^{(12)}$. Conditionally on the whole vector process Λ, the points of different classes occur in independent time-dependent Poisson processes.

Simple properties of the process are fairly easily calculated. A special case is that of stationary orderly processes with $\Lambda^{(1)}(t) = \Lambda^{(2)}(t)$; the two classes of points correspond to two independent realizations of the same time-dependent Poisson process. Then $\rho_1 = \rho_2$ and the four second-order intensity functions $h_b^{(a)}$ are the same, a fact that could be used to examine consistency with the model.

For cluster processes, bivariate processes can be formed in various ways. For instance, one may observe the cluster centres as class 1 points and the other points as class 2 points. Another possibility is to have two independent finite processes, one of class 1 points and the other of class 2 points, originating from each cluster centre; in other contexts, each cluster will consist of points of only one type, that class being determined at random. A few properties of the processes are outlined in Exercise 5.5.

For linear self-exciting processes, we simply replace (3.30) by

$$\rho^{(a)}(t; \mathscr{H}_t^{(1)}, \mathscr{H}_t^{(2)}) = \gamma_a + \int_{-\infty}^{t} w_{1a}(t - z)dN^{(1)}(z)$$

$$+ \int_{-\infty}^{t} w_{2a}(t - z)dN^{(2)}(z),$$

for $a = 1,2$. Hawkes (1971a, b, 1972) and Brillinger (1975) give a detailed discussion.

(iii) *Processes based on recurrence-times*

In discussing univariate point processes, we distinguished those in which the underlying structure depends in an essentially simple way on intervals between successive points. Equivalently the complete intensity function at time t is most simply specified in terms of the backward recurrence time $u(t)$, the interval from the preceding point to the one before that, and so on. The nature of the dependence on previous points should be contrasted with that in the processes in Section 5.3(ii). If the complete intensity function depends only on $u(t)$, then we have a renewal process.

For bivariate processes we can proceed similarly. We examine here only orderly processes in which the complete intensity functions at time t are determined by $(U^{(1)}(t), U^{(2)}(t))$, the pair of backward recurrence times. It is convenient to write $U(t) = \min \{U^{(1)}(t), U^{(2)}(t)\}$ for the backward recurrence time in the superposed process, i.e. the time from t back to the first point regardless of class; let $M(t)$ denote the class of that first point.

It is a mark of the richness of the bivariate theory that there are a considerable number of particular cases even within this already very special family. The following discussion is by no means exhaustive.

Unfortunately, while the specification via the complete intensity function exposes the structure of the process and suggests how to generate realizations of the process, the derivation of the main properties of the process is, in general, difficult. A natural first step in investigating the process is to find the equilibrium density $p(u_1, u_2)$ of the backward recurrence time $(U^{(1)}, U^{(2)})$; this exists only for suitable complete intensity functions $\rho^{(1)}(u_1, u_2)$ and $\rho^{(2)}(u_1, u_2)$. Realizations of the stochastic process of backward recurrence times consist of sections of deterministic increase at unit rate; when a point occurs the corresponding recurrence time drops to zero. This process is, by definition of the family, a Markov process. The equilibrium equations, obtained by considering the possible transitions in $(t, t + \delta)$, are

(Oakes, 1976)

$$(\partial/\partial u_1 + \partial/\partial u_2)p(u_1, u_2) = -\rho(u_1, u_2)p(u_1, u_2), \qquad (5.12)$$

$$p(0, u_2) = \int_0^\infty \rho^{(1)}(u_1, u_2)p(u_1, u_2)\,du_1,$$

$$p(u_1, 0) = \int_0^\infty \rho^{(2)}(u_1, u_2)p(u_1, u_2)\,du_2, \qquad (5.13)$$

where $\rho(u_1, u_2) = \rho^{(1)}(u_1, u_2) + \rho^{(2)}(u_1, u_2)$. The partial differential equation (5.12) has the solution

$$p(u_1, u_2) = \tilde{p}(u_1, u_2)g(u_1 - u_2),$$

where g is arbitrary and

$$\tilde{p}(u_1, u_2) = \begin{cases} \exp\left\{ -\int_0^{u_1} \rho(t, t + u_2 - u_1)\,dt \right\} & (u_1 \le u_2), \\ \exp\left\{ -\int_0^{u_2} \rho(t + u_1 - u_2, t)\,dt \right\} & (u_1 > u_2). \end{cases}$$

Equations (5.13) now form a pair of simultaneous integral equations most conveniently written in terms of $g_1(x) = g(-x)$ $(x \le 0)$ and $g_2(x) = g(x)$ $(x > 0)$.

Solutions are possible in the various special cases to be described below and also when

$$\rho^{(1)}(u_1, u_2) = \partial P(u_1, u_2)/\partial u_1, \quad \rho^{(2)}(u_1, u_2) = \partial P(u_1, u_2)/\partial u_2, \quad (5.14)$$

in which case $p(u_1, u_2) = e^{-P(u_1, u_2)}$ satisfies (5.12). The simplest example of this is to have

$$\rho^{(1)}(u_1, u_2) = a_{11}u_1 + a_{12}u_2 + b_1, \quad \rho^{(2)}(u_1, u_2) = a_{21}u_1 + a_{22}u_2 + b_2,$$

with $a_{12} = a_{21}$ and all a_{ij} and b_i non-negative. Provided that $a_{12}^2 \le 4a_{11}a_{22}$, the equilibrium distribution of (U_1, U_2) is truncated bivariate normal.

Some particular processes whose theory can be approached either via the above general results or from first principles are as follows:

(i) $\rho^{(a)}$ is a function only of $u_a(a = 1, 2)$, when the process is formed from two completely independent renewal processes;

(ii) both $\rho^{(1)}$ and $\rho^{(2)}$ are functions only of u_1, when the process is purely class 1 dependent and the class 1 points form a renewal process;

(iii) $\rho^{(1)}(u_1, u_2) = 0$ $(u_1 < u_2)$, $\rho^{(2)}(u_1, u_2) = 0$ $(u_2 < u_1)$, when the process is alternating, i.e. has deterministically alternating classes;

(iv) $\rho^{(1)}$ is a function only of u_2 and $\rho^{(2)}$ is a function only of u_1;

(v) both $\rho^{(1)}$ and $\rho^{(2)}$ are functions only of the backward recurrence time, $u(t)$, in the pooled process and its mark, $m(t)$, so that the complete intensity functions depend only on the last point.

The most important of these special cases is (v), which gives the two-state semi-Markov processes of Section 3.2(ii), in a thin disguise. Given that a point is of class a, the probability that it is followed by another class a point is, in a natural notation,

$$\int_0^\infty du \rho^{(a)}(u; M = a) \exp\left[- \int_0^u dv \{\rho^{(a)}(v; M = a) + \rho^{(b)}(v; M = a)\} \right],$$
(5.15)

and the density of the time between two points, given that they are both of class a, is the integrand of (5.15) divided by (5.15) itself as normalizing constant. Further, the Markov character of the mark sequence $\{M_j\}$ and the independence of the intervals given the sequence $\{M_j\}$ follow from the special form of the complete intensity functions. Conversely, given the four densities of interpoint intervals and the transition matrix of the two-state Markov chain $\{M_j\}$, the complete intensities are uniquely determined. The relevant formulae are essentially those occurring in the theory of competing risks (Chiang, 1968).

It is possible to solve (5.12) and (5.13) for this special case, although, in fact, important properties of the process can be calculated more directly from the conditional intensity functions $h_b^{(a)}$, obtained as in Section 3.2(ii). The arguments of that section are possible because each point of known class is a regeneration instant; in general, this is not the case for processes satisfying (5.12) and (5.13).

Berman (1978) introduced a general family of processes with a regenerative property. Suppose that class 1 points form a renewal process. Suppose further that in the interval between a pair of class 1 points, class 2 points are distributed in any way depending only on that interval and independent of occurrences in other intervals between class 1 points. Then class 1 points, but not in general class 2 points, are regeneration instants for the whole process.

(iv) *Processes produced by random displacements*
One simple way of obtaining a bivariate point process is to start with an arbitrary stationary point process of class 1 points and then to translate each class 1 point by a random amount to produce a class 2 point. We assume that the displacements are independent and identically distributed random variables with density f, that the displacements are independent of the originating point process, and

that the correspondence between individual class 1 and class 2 points is not observed.

One application is where the class 1 points are arrivals of customers in a queue and class 2 points are departures. If the number of servers is large enough for waiting to be negligible, the displacements are the service-times.

Some properties of the translated process of class 2 points considered as a univariate point process, have been studied in Section 4.4. It follows from the argument given there that

$$h_1^{(2)}(t) = f(t) + \int_{-\infty}^{\infty} h_1^{(1)}(u) f(t-u) \, du, \tag{5.16}$$

and, from (4.16), we have that

$$h_2^{(2)}(t) = \int_{-\infty}^{\infty} h_1^{(1)}(u) f_D(t-u) \, du, \tag{5.17}$$

where f_D is the convolution difference of f, i.e. the density of the difference D of two random variables independently distributed with density f. Note that, from (5.3), $h_2^{(1)}(t) = h_1^{(2)}(-t)$, and therefore (5.16) determines both cross-intensity functions. If, in particular, the class 1 process is a Poisson process of rate ρ_1, then

$$h_1^{(2)}(t) = f(t) + \rho_1; \tag{5.18}$$

it is easily shown that marginally the class 2 points also form a Poisson process.

The formulation sketched above appears to treat the class 1 and the class 2 points unsymmetrically. A generalization in which, incidentally, the two classes of point are treated more symmetrically is as follows: for simplicity we deal with processes based on the Poisson process. Consider an unobserved originating Poisson process of rate ρ_{12}, and two further 'noise' Poisson processes of rates ρ_1' and ρ_2', the three Poisson processes being mutually independent. Let the process of class 1 points be the superposition of a random translation of the originating Poisson process with the first 'noise' process, and the class 2 points be formed by an independent translation of the same originating process superposed with the second 'noise' process. We assume that the two sets of translations are determined by independent sequences of independent and identically distributed random variables with possibly different densities. In particular applications, one or both 'noise' processes may be absent. When both are absent we can equivalently regard, say, the class 1 points as fixed and the class 2 points as produced by translation with translation distribution the convolution difference of the above two densities.

5.4 An application to electronic counters

In the previous section we have deliberately concentrated on the formulation of processes and have given rather little detail about the derivation of properties. We now illustrate the use of bivariate point processes in an application to a counting problem in physics.

Suppose that there are two counters and three independent Poisson processes of points of rates ρ_1', ρ_2' and ρ_{12} respectively. Points from the first process are recorded only on counter 1. Points from the second process are recorded only on counter 2. Points from the third process are recorded virtually simultaneously on both counters. This situation is equivalent to one in which there is a single Poisson process of points of rate ρ, all of which would be recorded were the counters fully efficient. If the counters have efficiencies ε_1 and ε_2, then, subject to some obvious independence assumptions, we have the previous formulation with

$$\rho_1' = \rho\varepsilon_1(1 - \varepsilon_2), \quad \rho_2' = \rho\varepsilon_2(1 - \varepsilon_1), \quad \rho_{12} = \rho\varepsilon_1\varepsilon_2.$$

Now suppose that both counters are of type I with constant dead times τ_1 and τ_2; see Section 4.3 for discussion of a single counter. If dead times could be ignored, we would have a very simple bivariate point process, in fact that introduced at the end of Section 5.3(i).

Various properties of the process of recorded points may be of interest. Here we concentrate on finding the cross-intensity functions, from which, in particular, the covariance of the numbers of points recorded on the two counters in a given time can be found via (5.6) and (5.8). For simplicity we suppose that $\tau_1 = \tau_2 = \tau$, say.

The sequence of recorded points on one counter forms a renewal process, with an interval distribution that is exponential with origin at τ, i.e. each interval is the sum of the constant τ and a forward recurrence time in the relevant Poisson process; see Section 4.3 and Fig. 4.1. The first interval depends, of course, on the choice of origin. The equilibrium probability, p_a, that counter a is open is $\{1 + (\rho_a' + \rho_{12})\tau\}^{-1}$. To calculate $h_1^{(2)}$ for $t > 0$, we study the class 2 points, i.e. points recorded on counter 2, taking the origin at a class 1 point. Hence an essential step is to find the implication for counter 2 of the choice of origin.

For this, consider the 'state' of the two counters as a Markov process, the possible states of the system being as follows:

(i) both counters are open, with equilibrium probability p_{12};

(ii) counter 1 has been blocked for time u and counter 2 is open, with equilibrium probability density $q_1(u)$ for $0 \leq u \leq \tau$;

(iii) counter 1 is open and counter 2 has been blocked for time u, with equilibrium probability density $q_2(u)$ for $0 \leq u \leq \tau$;

(iv) counter 1 has been blocked for time u_1 and counter 2 has been blocked for time u_2, with equilibrium probability density $q_{12}(u_1, u_2)$, for $0 \leq u_1, u_2 \leq \tau$.

Because there is a non-zero probability of both counters becoming blocked simultaneously, leading to $u_1 = u_2$, it is convenient to isolate the singular component of the density in (iv) by writing

$$q_{12}(u_1, u_2) = q_{12}^{(s)}(u_1)\delta(u_1 - u_2) + q_{12}^{(c)}(u_1, u_2),$$

where $q_{12}^{(c)}(u_1, u_2)$ is absolutely continuous.

The equilibrium equations of the Markov process can now be obtained by examining transitions in a small interval $(t, t + \delta)$. The equations are, with $\rho_1 = \rho_1' + \rho_{12}, \rho_2 = \rho_2' + \rho_{12}$,

$$(\rho_1' + \rho_2' + \rho_{12})p_{12} = q_1(\tau) + q_2(\tau) + q_{12}^{(s)}(\tau),$$

$$q_1(0) = \rho_1' p_{12}, \quad q_2(0) = \rho_2' p_{12}, \quad q_{12}^{(s)}(0) = \rho_{12}p_{12}, \quad (5.19)$$

$$q_{12}^{(c)}(u, 0) = \rho_2 q_1(u), \quad q_{12}^{(c)}(0, u) = \rho_1 q_2(u);$$

$$\begin{aligned} q_1'(u) &= -\rho_2 q_1(u) + q_{12}^{(c)}(u, \tau), \\ q_2'(u) &= -\rho_1 q_2(u) + q_{12}^{(c)}(\tau, u); \end{aligned} \quad (5.20)$$

$$\partial q_{12}(u_1, u_2)/\partial u_1 + \partial q_{12}(u_1, u_2)/\partial u_2 = 0. \quad (5.21)$$

Finally, there is a normalizing condition that the total probability is unity.

This system is solved by noting from (5.21) that $q_{12}(u_1, u_2)$ is a function of $u_1 - u_2$ and, therefore, in particular, that $q_{12}^{(s)}(u) = q_{12}$, a constant. On using the last part of (5.19) in (5.20), we have that

$$\begin{aligned} q_1'(u) + \rho_2 q_1(u) &= \rho_1 q_2(\tau - u), \\ q_2'(u) + \rho_1 q_2(u) &= \rho_2 q_1(\tau - u). \end{aligned} \quad (5.22)$$

This being a linear system, we look for exponential solutions. It is easily shown by substitution that, provided that $\rho_1 \neq \rho_2$, (5.22) has the general solution

$$\begin{aligned} q_1(u) &= A\rho_1 + Be^{-\rho_1 \tau}e^{(\rho_1 - \rho_2)u}, \\ q_2(u) &= A\rho_2 + Be^{-\rho_2 \tau}e^{(\rho_2 - \rho_1)u}, \end{aligned}$$

where A and B are constants. The substitution of these functions into (5.19) and the use of the normalizing condition determines A and B

and proves that

$$p_{12} = \frac{p_1 p_2 (\rho_1 e^{\rho_1 \tau} - \rho_2 e^{\rho_2 \tau})}{(\rho_1' e^{\rho_1 \tau} - \rho_2' e^{\rho_2 \tau})}. \tag{5.23}$$

The next step is to introduce the choice of time origin at an instant at which a point is recorded on counter 1, implying in particular that counter 1 is open at time $0-$. Therefore, the state of counter 2 at time $0-$ has the following distribution:

(a) open, with probability p_{12}/p_1;

(b) blocked, having been blocked for time u, with probability density $q_2(u)/p_1$.

In case (b), counter 2 is unchanged at time $0+$, but for case (a) there are two possibilities, depending on the nature of the originating point on counter 1. Thus at time $0+$, we have three possibilities for the state of counter 2, namely

(a)$'$ just beginning a blocked period, with probability $(p_{12}\rho_{12})/(p_1\rho_1)$;

(a)$''$ open, with probability $(p_{12}\rho_1')/(p_1\rho_1)$;

(b) as above.

For (a)$'$, (a)$''$ and (b), the class 2 points form a renewal process with a displaced exponential distribution for the intervals, and with different starting intervals in the three cases. Case (a)$'$ leads to an ordinary renewal process with renewal density $h^{(2)}(t)$ having Laplace transform $\rho_2\{(\rho_2 + s)e^{s\tau} - \rho_2\}^{-1}$, while case (a)$''$ leads to a modified renewal process the first interval of which has an exponential distribution with parameter ρ_2. Case (b) gives an ordinary renewal process starting from time $-u$. It follows that

$$h_1^{(2)}(t) = \frac{p_{12}\rho_{12}}{p_1\rho_1} h_2^{(2)}(t) + \frac{p_{12}\rho_1'}{p_1\rho_1}\left\{\rho_2 e^{-\rho_2 t} + \int_0^t \rho_2 e^{-\rho_2 y} h_2^{(2)}(t-y)\,dy\right\}$$

$$+ \frac{1}{p_1}\int_0^\tau q_2(u) h_2^{(2)}(t+u)\,du.$$

Therefore, taking Laplace transforms, we have that

$$h_1^{(2)*}(s) = \frac{p_{12}\rho_1'\rho_2}{p_1\rho_1(\rho_2 + s)} + \frac{\rho_2}{(\rho_2 + s)e^{s\tau} - \rho_2}\left[\frac{p_{12}\rho_{12}}{p_1\rho_1} + \frac{p_{12}\rho_1'\rho_2}{p_1\rho_1(\rho_2 + s)}\right.$$

$$\left. + p_2\left\{\rho_2\frac{e^{s\tau}-1}{s} + \frac{p_{12}(\rho_2-\rho_1)e^{\rho_1\tau}}{\rho_1' e^{\rho_1\tau} - \rho_2' e^{\rho_2\tau}}\frac{e^{(s-\rho_1+\rho_2)\tau}-1}{s-\rho_1+\rho_2}\right\}\right]. \tag{5.24}$$

There is an analogous expression for $h_2^{(1)}(t)$.

The final step is to extract useful information from (5.24) or its equivalent. For example, it is easily shown from (5.8) that the Laplace transform of the covariance, $C(t)$, of $N^{(1)}(t)$ and $N^{(2)}(t)$ is

$$\mathcal{L}\{C(t); s\} = \rho_1 p_1 h_1^{(2)*}(s)/s^2 + \rho_2 p_2 h_2^{(1)*}(s)/s^2$$
$$+ \rho_{12} p_{12}/s^2 - 2\rho_1 \rho_2 p_1 p_2/s^3.$$

The substitution of the special form (5.24) leads to

$$\mathcal{L}\{C(t); s\} = \rho_{12} p_1 p_2 p_{12}/s^2 + l/s + c_0^*(s),$$

where l is a constant whose value can be calculated and $c_0^*(s)$ is an analytic function of s in a half plane of the form $\text{Re}(s) > -\gamma_0$ for some $\gamma_0 > 0$. It follows that as $t \to \infty$, $C(t)$ is approximated by

$$\rho_{12} p_1 p_2 p_{12} t + l, \tag{5.25}$$

with an error that is exponentially small in t. Of course this is only one aspect of the bivariate point process that can be investigated via the results given here.

We have studied this particular process in a little detail to illustrate the mathematical arguments that may be involved in deriving explicit properties of special processes.

5.5 Marked point processes

As pointed out in Section 5.2(ii), one can specify a bivariate or multivariate point process in terms of a univariate process of points, together with an indicator sequence of random variables giving the class of each point. This suggests the generalization in which a random variable, to be called a mark, is attached to each point and where the mark is not restricted to be an indicator of point class. For example, the mark may be normally distributed or, in another interpretation, may, in a non-orderly process, give the number of points at each instant of occurrence; see Section 1.4(i).

Two applications involving continuous marks are in neurophysiology, where the magnitude of an electrical pulse may be specified as well as its time of occurrence, and in particle physics, where the energy, and possibly momentum, of a particle may be specified as well as its time of origin.

It is, of course, stultifying to draw rigid boundaries between different subjects. If, however, attention is focused strongly on the process of marks, and rather weakly on the process of points, it may be convenient to treat the process via techniques not primarily those of point processes.

If we denote the mark of the ith point by M_i and work from a convenient origin, we can write

$$M(t) = \sum_{\{i\,:\,0\,<\,T_i\,\leq\,t\}} M_i = \int_0^t M_{\{u\}}\,dN(u) = \int_0^t dM(u), \quad (5.26)$$

where $M_{\{u\}}$ is the mark of a point occurring at u and $M(t)$ is the sum of marks; in general we write $M(A)$ for the sum of the marks over the points in A.

One natural set of questions concerns the stochastic process M and in particular the joint distribution of $(M(t), N(t))$. We now distinguish sharply two types of situation. In the first the marks are independent and identically distributed random variables independent of the point process. In the second the object of study is the dependence between the points in the originating point process and the marks.

Suppose that we have an orderly univariate point process of rate ρ and conditional intensity function h. Suppose that the marks are independent and identically distributed random variables with density f_M. The argument of Section 1.4(i) generalizes immediately. Given $N(A) = n$, $M(A)$ is the sum of n independent and identically distributed random variables, so that

$$E\{e^{-sM(A)}|N(A) = n\} = \{f_M^*(s)\}^n. \quad (5.27)$$

Hence the joint Laplace transform of $M(A)$ and probability generating function of $N(A)$ is

$$E\{e^{-sM(A)}z^{N(A)}\} = G_{N(A)}\{zf_M^*(s)\}. \quad (5.28)$$

Asymptotic bivariate normality of $(M(A), N(A))$ will occur under weak conditions. It follows from the conditional mean and variance implicit in (5.27) that

$$E\{M(A)\} = \rho|A|E(M), \quad (5.29)$$

$$\text{var}\{M(A)\} = \rho|A|\,\text{var}(M) + \{E(M)\}^2\,\text{var}\{N(A)\}, \quad (5.30)$$

$$\text{cov}\{M(A), N(A)\} = E(M)\,\text{var}\{N(A)\}.$$

In the special case when A is the interval $(0, t]$, we can, of course, use (2.27) for $\text{var}\{N(A)\}$.

In many ways, however, the more interesting situation is where the marks may depend on one another and on the point process. Denote by $\mathscr{H}_t^{\Pi}$ the history of the unmarked point process at time t and by $\mathscr{H}_t^M$ the history of the mark sequence, i.e. the sequence of marks at points at or before t without specification of the time instants. Then $(\mathscr{H}_t^{\Pi}, \mathscr{H}_t^M)$ is the history of the marked point process at t. We now

Table 5.1. *Some special types of marked point process*

Dependence of ρ	Dependence of f_M		
	None	*On $\mathcal{H}^\Pi$*	*On $\mathcal{H}^M$*
None	1. Poisson process with random marks	2. Poisson process with dependent marks	3. Poisson process with autodependent marks
On $\mathcal{H}^\Pi$	4. Point process with random marks	5. Point process with dependent marks	6. Point process with autodependent marks
On $\mathcal{H}^M$	7. Point process controlled by random marks	8. —	9. Point process controlled by autodependent marks

$\mathcal{H}^\Pi$: history of point process;
$\mathcal{H}^M$: history of marks;
ρ: complete intensity of point process;
f_M: density of mark.

consider the process as specified by the complete intensity function $\rho(t; \mathcal{H}_t^\Pi, \mathcal{H}_t^M)$ of the point process at t and by $f_M(m; t, \mathcal{H}_t^\Pi, \mathcal{H}_t^M)$, the density of a mark at t given a point there and given the complete history $(\mathcal{H}_t^\Pi, \mathcal{H}_t^M)$.

In applications there are, of course, many special cases of possible interest. A number are summarized in Table 5.1. We comment here only on a few of the nine possibilities listed. Our emphasis is on processes with a simple description. Problems of identifiability will not be discussed; that is, some processes may appear in Table 5.1 in more than one place.

Possibilities 1 and 4, in which a completely random process of marks is attached to a given point process, have already been studied. Case 5 is a natural representation when the marks are, in some reasonable sense, dependent on the point process. An important special case is where the point process is a renewal process and the mark at a particular point depends only on the interval back to the preceding point. The process is then simply described in terms of X_n, the interval from the $(n-1)$st to the nth point and M_n, the mark at the nth point. In fact, (X_n, M_n) form independent and identically distributed bivariate random variables. It follows in particular that, subject to the existence of variances,

$$(X_1, M_1) + \ldots + (X_n, M_n) \tag{5.31}$$

is asymptotically bivariate normal and properties of $(N(t), M(t))$
follow by the argument of (2.33).

This process is very similar to, but to be sharply distinguished from,
the corresponding special case of possibility 7. Here the 'direction of
causality' is reversed. Suppose that marks are independent and
identically distributed random variables and that the interval X_{n+1}
from the nth to the $(n+1)$st point depends only on M_n. Then the point
process is clearly a renewal process and the pairs (X_{n+1}, M_n) are
independent. Arguments similar to those outlined in connection with
(5.31) give properties of $(N(t), M(t))$. For large t the behaviour is the
same, although the local properties of the processes are reversed.

Unfortunately, not all simple situations are easily captured within
the framework of Table 5.1. For example, suppose that a stationary
random process is sampled at the points of a point process, i.e. the
mark at a particular point is the value of the stationary process at that
point. Then, even if the point process and the stationary process are
both of simple structure, the marked point process is likely to show
dependencies on $(\mathcal{H}^{\Pi}, \mathcal{H}^M)$ and so not be one of the simple types of
Table 5.1. Thus if an Ornstein–Uhlenbeck process is sampled by the
points of an independent Poisson process, the density of the mark at a
particular point depends on both the previous mark and on the
interval back to the previous sampling point.

Some more complex marked processes will be discussed in Section
5.6; see also Section 2.4 in connection with Palm measures.

5.6 More complex marked processes

(i) Introduction

In the previous section it was assumed that the marks are arbitrary
random variables. In most of the special examples one-dimensional
random variables have been considered, although the generalization
to multidimensional marks is immediate. A more significant generalization is to allows the mark to be not just a single value or vector of
values but to be a function of time, in general a random function of
time. There then results a new derived stochastic process, obtained by
summing the marks associated with all the points.

Thus if associated with a point at u there is a random function of
time $M_{\{u\}}(t - u)$, i.e. with the 'origin' at u, we can define the stochastic
process

$$Z(t) = \int_{-\infty}^{\infty} M_{\{u\}}(t - u)\, dN(u) = \sum_i M_{\{T_i\}}(t - T_i).$$

In all the applications we shall consider, the random mark functions are one-sided, i.e. zero for negative arguments. Then

$$Z(t) = \int_{-\infty}^{t} M_{\{u\}}(t - u)\, dN(u) = \sum_{\{i\,:\,T_i \leq t\}} M_{\{T_i\}}(t - T_i). \qquad (5.32)$$

It is convenient to begin by discussing some special cases.

(ii) *Simple shot noise*

Suppose that the generating point process is a Poisson process of rate ρ and that the marks are exponential functions with random 'amplitudes' and a constant decay parameter, i.e.

$$M_{\{T_i\}}(t) = M_i e^{-\kappa t} \qquad (t > 0),$$

where the $\{M_i\}$ are independent and identically distributed random variables independent of the point process. Fig. 5.1 shows a typical realization. We call the resulting process

$$Z(t) = \sum_{\{i\,:\,T_i \leq t\}} M_i e^{-\kappa(t - T_i)} \qquad (5.33)$$

a simple shot noise process. In a special case the M_i are constant.

This process was first considered in connection with the shot effect: electrons arrive at an electrode of a valve and each arrival generates a

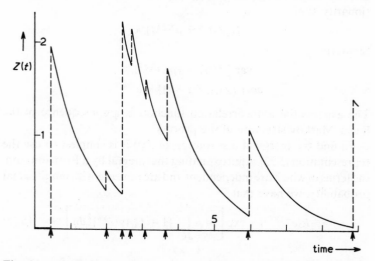

Fig. 5.1. *Shot-noise process.* ↑, *points of underlying point process. Process jumps upwards a random amount at each point and decays exponentially between points.*

signal (voltage or current). The process has also been suggested in hydrology as a rough model for river flows: here the points represent storms (i.e. input into the system). A different interpretation is obtained by imagining the time axis reversed and supposing that a point at T_i corresponds to a future point income of amount M_i, which discounted at compound interest of rate κ, has present value at time zero of $M_i e^{-\kappa T_i}$. It then follows that $Z(t)$ is the value at time t of all 'future' income discounted to time t.

The process (5.33) is related to the simpler marked process of Section 5.5, but takes a weighted sum of marks more complicated than an unweighted sum.

The theory of (5.33) can be approached in several ways. One, not in the spirit of point processes, is to note that (5.33) is a form of first-order autoregressive process, being a Markov process in continuous time with, for $\delta > 0$,

$$Z(t + \delta) = e^{-\kappa\delta} Z(t) + \varepsilon'_\delta(t),$$

where $\varepsilon'_\delta(t)$ is an 'innovation' independent of $Z(t)$. For small δ

$$Z(t + \delta) = (1 - \kappa\delta)Z(t) + \varepsilon_\delta(t) + o(\delta), \tag{5.34}$$

where $\varepsilon_\delta(t)$ is zero with probability $1 - \rho\delta + o(\delta)$ and with probability $\rho\delta + o(\delta)$ has the distribution of a random variable M_i. It follows directly from (5.34), on taking expectations and assuming stationarity, that

$$E\{Z(t)\} = \rho E(M)/\kappa.$$

Similarly

$$\mathrm{var}\,\{Z(t)\} = \tfrac{1}{2}\rho E(M^2)/\kappa,$$
$$\mathrm{corr}\,\{Z(t), Z(t + \tau)\} = e^{-\kappa|\tau|}.$$

The exponential autocorrelation function is a consequence of the linear Markov structure of the process.

To find the marginal distribution of $Z(t)$ it is simplest to use the representation (5.32). Approximating the integral by a Riemann sum, the terms of which are independent and are nonzero with infinitesimal probability, we have that

$$E(e^{-sZ(t)}) = \exp\left[-\rho \int_0^\infty \{1 - f_M^*(se^{-\kappa u})\}\,du \right]. \tag{5.35}$$

(iii) *Marks of random extent*
We continue to assume that the underlying point process is a Poisson process of rate ρ. Suppose that the mark attached to an arbitrary

point is a rectangle of unit height and random base, i.e.

$$M_{\{T_i\}}(t) = \begin{cases} 1 & (0 \leq t \leq L_i), \\ 0 & (L_i < t), \end{cases}$$

where the $\{L_i\}$ are independent nonnegative random variables, identically distributed with density f_L, the sequence being independent also of the point process. We can consider T_i as the left-end of an individual of extent L_i. Then $Z(t)$ is the number of individuals present at time t. Fig. 5.2 illustrates the process.

We can give a number of closely related interpretations of the stochastic process Z. It represents an immigration-death process in which new individuals enter in a Poisson process of rate ρ and in which f_L specifies the distribution of the life-time of an individual. It represents a queueing system in which the points are arrivals of customers, in which constraints on service are negligible and in which f_L gives the distribution of service-time. It represents a textile yarn in which the points are fibre left-ends and in which f_L gives the distribution of fibre length. In all these cases $Z(t)$ is the number of individuals present at time t.

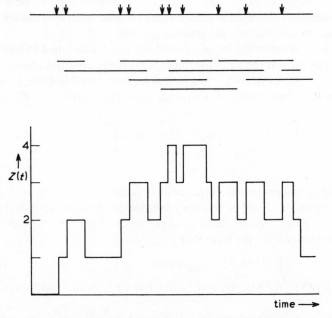

Fig. 5.2. *Special form of marked process.* ↓, *points of underlying point process. Each point defines line of random extent as shown, vertical position of lines being irrelevant. Z(t), number of lines covering t.*

If f_L is exponential, the process Z is a Markovian immigration-death process, whose properties can be studied via the Kolmogorov forward or backward equations (Cox and Miller, 1965, p. 168). The equilibrium distribution of $Z(t)$ is Poisson with mean $\rho E(L)$. If L is constant, l say, then $Z(t)$ simply counts the number of points in $(t - l, t)$ and has a Poisson distribution of mean ρl. Further, the joint distribution of $Z(t)$ and $Z(t + \tau)$ is bivariate Poisson with $\operatorname{cov}\{Z(t), Z(t + \tau)\} = \rho \max(l - \tau, 0)$. The process is in a reasonable sense a simple moving average of a Poisson process.

The fact that $Z(t)$ has a marginal Poisson distribution both when L is constant and when L is exponentially distributed suggests that the conclusion holds for arbitrary distributions of L. One way of proving this is to suppose first that the distribution of L is discrete with probabilities $\{p_i\}$ attaching to values $\{l_i\}$. Then points with mark of extent l_i arise in a Poisson process of rate ρp_i; moreover the different types of mark occur independently. Thus $Z(t) = \Sigma Z_i(t)$, where $Z_i(t)$ is the contribution from marks of extent l_i. The argument for marks of constant extent shows that $Z_i(t)$ has a Poisson distribution of mean $\rho p_i l_i$ and, because of independence, it follows that $Z(t)$ has a Poisson distribution of mean $\Sigma \rho p_i l_i = \rho E(L)$, the required result. Because a continuous distribution can be approximated arbitrarily closely by a discrete distribution, the result is general.

The joint distribution of $Z(t)$ and $Z(t + \tau)$ can be found by a similar argument, but the autocovariance is probably obtained most simply from (5.32) by introducing a formal random variable $dN(u, l)$ for the number of points in $(u, u + du)$ with associated extent $(l, l + dl)$, so that

$$Z(t) = \int_{l=0}^{\infty} \int_{u=t-l}^{t} dN(u, l), \quad Z(t + \tau) = \int_{m=0}^{\infty} \int_{v=t+\tau-m}^{t+\tau} dN(v, m).$$

$$(5.36)$$

The formal properties of the random variables $\{dN(u, l)\}$, analogous to those of $\{dN(u)\}$ in Section 2.5, follow from the Poisson character of the point process and the independence of the marks attached to different points. We have that

$$E\{dN(u, l)\} = \rho f_L(l) \, du \, dl, \tag{5.37}$$

$$E\{dN(u, l) \, dN(v, m)\} = \{\rho^2 f_L(l) f_L(m) + \rho f_L(l) \delta(u - v) \delta(l - m)\}$$
$$\times \, du \, dv \, dl \, dm,$$

$$\operatorname{cov}\{dN(u, l), dN(v, m)\} = \rho f_L(l) \delta(u - v) \delta(l - m) \, du \, dv \, dl \, dm. \tag{5.38}$$

Use of (5.36) and (5.37) yields another proof that $E\{Z(t)\} = \rho E(L)$ and

(5.36) and (5.38) give that, for $\tau \geq 0$,

$$\text{cov}\{Z(t), Z(t+\tau)\} =$$

$$\int_{l=0}^{\infty} \int_{m=0}^{\infty} \int_{u=t-l}^{t} \int_{v=t+\tau-m}^{t+\tau} \text{cov}\{dN(u, l), dN(v, m)\}$$

$$= \rho \int_{\tau}^{\infty} (l-\tau) f_L(l)\, dl$$

$$= \rho \int_{\tau}^{\infty} \mathscr{F}_L(l)\, dl, \qquad (5.39)$$

where $\mathscr{F}_L$ is the survivor function corresponding to the density f_L.

One consequence of (5.39) is that the distribution of L can be deduced from the autocorrelation function of Z. The results for constant L and for exponentially distributed L conform with those obtained by the alternative arguments sketched above.

Generalization of (5.37) and (5.38) provides the easiest way of obtaining first and second moment properties in more complicated situations.

(iv) *Some general second-order results*
In Sections 5.6(ii) and (iii) we have discussed two important but very special processes. They are based on a Poisson process, they have mark functions of very special kinds and these mark functions are independent of the generating Poisson process. To some extent all these aspects can be generalized, especially if we concentrate on first- and second-moment properties assuming stationarity.

We regard the derived process as defined by (5.32) and investigate its properties via the formal relations

$$E\{M_{\{u\}}(x)\, dN(u)\} = \rho \mu_M(x)\, du, \qquad (5.40)$$

$$E\{M_{\{u\}}(x)\, dN(u) M_{\{v\}}(y)\, dN(v)\} = \rho h(v-u)\mu_M(x, y; v-u)\, du\, dv,$$

$$+ \rho \delta(v-u)\mu_M(x, y; 0)\, du\, dv, \quad (5.41)$$

where $\mu_M(x, y; \tau) = E\{M_{\{0\}}(x)M_{\{\tau\}}(y)\}$; see Section 2.5 for a similar argument. In (5.40), $\mu_M(x)$ gives the expected mark function a time x after the origin and in (5.41), the dependence of $\mu_M(x, y; v-u)$ on the third argument allows some dependence between the mark functions and the separation of the originating points. If the mark functions are independent of one another and of the point process

$$\mu_M(x, y; v - u) = \begin{cases} \mu_M(x)\mu_M(y) & (v \neq u), \\ \mu_M(x, y) & (v = u), \end{cases} \tag{5.42}$$

where $\mu_M(x, y)$ refers to a single mark function.

It follows that

$$E\{Z(t)\} = \rho \int_0^\infty \mu_M(x)\,dx, \tag{5.43}$$

$$E\{Z(t)Z(t + \tau)\} = \rho \int_0^\infty \int_0^\infty dx\,dy\,h(\tau + x - y)\mu_M(x, y; \tau + x - y)$$

$$+ \rho \int_0^\infty dx\,\mu_M(x, \tau + x; 0), \tag{5.44}$$

from which the main results of Sections 5.6(ii) and (iii) can be recovered as special cases.

As a final special case, consider the shot-noise process based on an arbitrary second-order stationary point process of rate ρ; the mark functions have the previous form of an exponential function of random amplitude. Then $\mu_M(x) = E(M)e^{-\kappa x}$, $E\{Z(t)\} = \rho E(M)/\kappa$ and after some calculation with (5.44), we have that

$$E\{Z(t)Z(t + \tau)\} = \frac{\rho}{2\kappa}\{E(M)\}^2 \int_{-\infty}^\infty e^{-\kappa|u|}h(\tau + u)\,du + \frac{\rho E(M^2)e^{-\kappa\tau}}{2\kappa}. \tag{5.45}$$

Bibliographic notes, 5

Multivariate point processes have a long history expecially in connection with applications in physics (Macchi, 1979; Srinivasan, 1969). A systematic account was given by Cox and Lewis (1972); the present account is largely based on that. Oakes (1976) studied Markov processes of intervals. Berman (1978) introduced a general family of regenerative point processes. Berman (1977), Daley and Milne (1975) and Wisniewski (1972) have examined Palm–Khintchine relations for multivariate point processes.

For further details of the counting problem of Section 5.4, see Cox and Isham (1977) and Kingman (1977).

Marked point processes are strongly emphasized by Matthes, Kerstan and Mecke (1978). Section 5.6 is largely based on Cox and Miller (1965, Chapter 9).

Further results and exercises, 5

5.1.   Construct a bivariate stationary point process in which the asymptotic correlation between $N^{(1)}(t)$ and $N^{(2)}(kt)$, as $t \to \infty$, is non-zero, where k is a constant different from one. [Section 5.2(i).]

5.2.   Discuss the extension of (5.2) to non-orderly processes. Note the possibilities of (a) retaining the definition (5.1) for $h_b^{(a)}$; (b) studying the expected number of points in $(t, t + \delta_2)$; (c) making the initial condition in $(-\delta_1, 0)$ take account of multiplicity. [Section 5.2(iii).]

5.3.   Suppose that class 1 points occur in a Poisson process of rate ρ_1 and that, conditionally on the class 1 points, the class 2 points occur in a time-dependent Poisson process of rate $g(u_t^{(1)})$, where g is a given function and $u_t^{(1)}$ is the backward recurrence time from t to the immediately preceding class 1 point. Prove that the rate ρ_2 of the class 2 process is given by $\rho_2 = \rho_1 g^*(\rho_1)$ and that the conditional intensity $h_2^{(2)}(\tau)$ satisfies

$$h_2^{(2)}(\tau) = \int_0^\tau \rho_1 e^{-\rho_1 u} g(u) du + \rho_2^{-1} \rho_1 e^{-\rho_1 \tau} \int_0^\infty g(u) g(u + \tau) e^{-\rho_1 u} du.$$

Obtain also the cross-intensity function. Suggest how the function g could be determined from a large amount of data (a) on class 2 points only, (b) on the whole process. For (b) suggest also how conformity with the model could be examined [Section 5.3(i).]

5.4.   A bivariate doubly stochastic Poisson process has a rate function with components

$$\Lambda^{(1)}(t) = \rho_1 + R_1 \cos(\omega_0 t + \theta + \Phi), \quad \Lambda^{(2)}(t) = \rho_2 + R_2 \cos(\omega_0 t + \Phi),$$

where ρ_1, ρ_2, ω_0 and θ are constants, Φ is uniformly distributed over $(0, 2\pi)$ independently of (R_1, R_2) and

$$E(R_1) = E(R_2) = 0, \ E(R_1^2) = \sigma_1^2, \ E(R_1 R_2) = \sigma_{12}, \ E(R_2^2) = \sigma_2^2;$$

the distribution of (R_1, R_2) is assumed such that $|R_i| \le \rho_i$. Show that the point process is stationary (but not ergodic), calculate the intensity and cross-intensity function and the corresponding spectral functions. [Sections 5.3(ii), 3.3(iii); Cox and Lewis (1972).]

5.5.   Consider a Neyman–Scott cluster process with the notation of Section 3.4. Suppose that cluster centres are observed as class 1 points and the derived cluster points as class 2 points. Prove that the cross-intensity function is

$$h_1^{(2)}(u) = E(M)f(u) + \rho_c E(M).$$

Suppose next that the cluster centres are not observed and that whole clusters are allocated at random to class 1 and to class 2. Show that the class 1 and class 2 processes are independent Neyman–Scott processes.

Thirdly, suppose that the cluster centres are unobserved but that with each cluster centre is associated independently a cluster of class 1 points, defined by $M^{(1)}$ and $f^{(1)}$, and a cluster of class 2 points defined by $M^{(2)}$ and $f^{(2)}$. Prove that

$$h_1^{(2)}(u) = E(M^{(2)}) \int_{-\infty}^{\infty} f^{(1)}(x) f^{(2)}(x + u)\, dx + \rho_c E(M^{(2)}).$$

Finally use the results of Section 5.3 to calculate the function $h_1^{(2)}(u)$ when the points of a univariate Neyman–Scott process are attached randomly to class 1 or 2. [Sections 5.3(ii) and 3.4.]

5.6.   Customers arrive in a Poisson process of rate α in a queueing system with two servers. If both servers are busy the customer is lost. Otherwise the customer will go to server I, unless server I is busy in which case the customer will go to server II. Let the service-time be constant, equal to a, say. Show that if a bivariate point process is formed from the instants of arrival at the two servers then the conditions for the special Markov process of intervals of Section 5.3(iii) are satisfied and that the equation of the process can be solved. [Section 5.3(iii); Oakes (1976).]

5.7.   Construct and study a simple example of a process in Section 8 of Table 5.1. [Section 5.5.]

CHAPTER 6

Spatial processes

6.1 Preliminary remarks

In the previous chapters we have discussed point processes occurring in time. We now turn more briefly to processes in which the points are distributed in space. Examples include plants of a particular species, idealized as points in a field; cities, idealized as points in a geographical area; cars idealized, at a particular time instant, as points along a road; galaxies, idealized as points in three-dimensional space, and so on.

There are two rather different features distinguishing temporal from spatial processes. First, time is directional, whereas for a spatial process in one dimension it will often be required to treat the two directions 'left to right' and 'right to left' symmetrically. For example, in constructing a model for the occurrence along a line transect of plants of a particular species, an emphasis on say 'left to right' in the model definition would be unwise biologically even if formally allowable mathematically. Thus our quite strong emphasis on the complete intensity function for temporal processes has to be abandoned.

Secondly, many of the spatial processes that we shall consider occur not in one dimension, e.g. along a line or on the circumference of a circle, but rather in some higher dimensional space. For simplicity, we consider mostly processes in some region of two-dimensional Euclidean space. Now in one dimension, not only can we define intervals between successive points, but the point process is in effect determined by the process of such intervals, plus appropriate initial conditions. In more than one dimension, we can define in an obvious way, for a given point, its nearest neighbour, and the nearest neighbour distance and direction, the second nearest neighbour, etc. These are indeed properties of considerable interest in some contexts. Nevertheless the spatial process is not determined by the set of nearest neighbour distances alone and the specification of a process via the complete set of distances and directions is in most cases unnatural.

Thus model formulation via the process of intervals has to be abandoned, at least in simple form.

These remarks show that while to some extent the mathematical formalism of the previous chapters carries over to spatial processes with little difficulty, there is a sense in which model formulation is much more difficult for spatial processes than for temporal processes. In fact, most of the special one-dimensional processes of Chapter 3 have been formulated in terms of intervals or the complete intensity function.

The situation is rather easier conceptually, if not mathematically, for spatial-temporal processes. That is, we consider processes of points distributed in space and time. There are two distinct categories of such processes. The first covers processes of points in space–time and an example is to define a point to be the place and instant of identification of a case of an infectious disease. The second category covers processes of points in space where the points continue for a while in time and these processes may be of interest in various ways. First the points may, in fact, have velocities, i.e. be moving in space. An example is that of cars distributed along a road. While such a process can be considered, at a particular time instant, as a purely spatial process, it is more interesting to consider it simultaneously as a temporal process. Secondly, the points may be fixed in space but new points may arise and old points disappear. Also, for either of the types of processes just described, we may look for a stationary or equilibrium distribution in time, as a model for a spatial process observed at one time instant. Note in this connection that to simulate a spatial process we will nearly always need a construction defined in time, so that study of a spatial process totally divorced from temporal considerations is in a sense artificial.

A new aspect of spatial processes is the possibility of studying geometrical properties of the pattern of points. For example, the collection of Dirichlet (or Voronoi) cells may be of interest: each such cell contains all elements in the plane having a particular point of the process as nearest point. Fig. 6.1 illustrates the construction. Because each point generates one cell, the mean area of a cell is ρ^{-1}; calculation of, for example, the distribution of cell areas is difficult.

For spatial processes, the simpler definitions and associated functions, such as the second-order intensity function, carry over from the earlier chapters in an immediate way and we shall not state any such definitions explicitly.

In Sections 6.2 and 6.3 some particular spatial processes are described, while in Section 6.4 a more general discussion of spatial

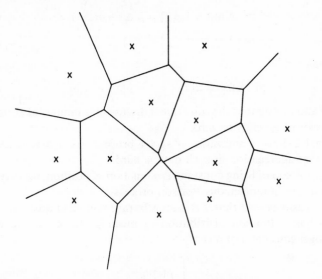

Fig. 6.1. *Dirichlet cells in plane,* ×, *point of defining point process. A cell consists of elements of plane closest to a particular point.*

processes is given. Finally, in Section 6.5, spatial–temporal processes are considered.

6.2 Some simple generalizations of one-dimensional processes

We now discuss briefly four processes whose definitions carry over fairly directly from one dimension. These are the Poisson process, the doubly stochastic Poisson process, the Poisson cluster process and the process with independent locations discussed in Section 3.2(vi). Finally, we mention the difficulty of generalizing the idea of a renewal process to a spatial situation. For simplicity, we deal almost entirely with processes in two dimensions; we denote the spatial position of a point by the vector ζ with components (ξ, η) relative to Cartesian axes in the plane.

(i) *Poisson processes*

We define a Poisson process of rate ρ by a direct generalization of the intensity definition (1.1)–(1.3). Let ρ be a positive constant with dimension $[\text{area}]^{-1}$. Let δ_ζ be a small neighbourhood of ζ of area $|\delta_\zeta|$ and denote by $\mathscr{E}_{\delta_\zeta}$ all points of the process occurring outside δ_ζ. Then if $N(B)$ denotes the number of points in a region B, we require that

$$\text{pr}\,\{N(\delta_\zeta) = 1 | \mathscr{E}_{\delta_\zeta}\} = \rho|\delta_\zeta| + o(|\delta_\zeta|), \tag{6.1}$$

$$\text{pr}\,\{N(\delta_\zeta) > 1 | \mathscr{E}_{\delta_\zeta}\} = o(|\delta_\zeta|), \tag{6.2}$$

so that

$$\text{pr}\,\{N(\delta_\zeta) = 0 | \mathscr{E}_{\delta_\zeta}\} = 1 - \rho|\delta_\zeta| + o(|\delta_\zeta|). \tag{6.3}$$

Here and throughout the subsequent discussion a limit is taken as the diameter of δ_ζ tends to zero.

Fig. 6.2 shows a realization of such a process. Visual assessment of apparent aggregation in such data is hard.

If ρ, instead of being constant, is a function of position, $\rho(\zeta)$ say, we have a non-homogeneous Poisson process.

It is easily shown that if B is an arbitrary region, of area $|B|$, then $N(B)$ has a Poisson distribution of mean $\rho|B|$, or, in the non-homogeneous case, of mean

$$\int_B \rho(\zeta)\,d\zeta.$$

Further, the numbers of points in non-overlapping regions are mutually independent.

Suppose now that an origin is taken either arbitrarily in the plane or at an arbitrary point of the process. Let R_1 be the distance from the origin to the nearest point. Then $R_1 > r$ if and only if a disc of area πr^2 contains no point, other than, perhaps, a defining point at the origin.

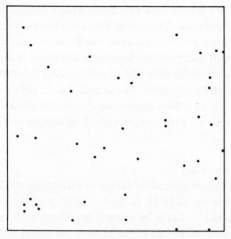

Fig. 6.2. *Realization of two-dimensional Poisson process.*

Thus

$$\operatorname{pr}(R_1 > r) = e^{-\rho\pi r^2},$$

so that, on differentiating, the probability density function of R_1 is

$$f_{R_1}(r) = 2\rho\pi r e^{-\rho\pi r^2}. \tag{6.4}$$

We call this the density of nearest neighbour distances. A more direct form of this result, implicit in the derivation, is that the area, $A_1 = \pi R_1^2$, covered by a disc centred on the origin and extending to the nearest point, is exponentially distributed with parameter ρ. More generally if $A_k = \pi R_k^2$ is the area to the kth nearest point, then the areas $A_1, A_2 - A_1, A_3 - A_2, \ldots$ are independent and exponentially distributed with parameter ρ. Thus A_k and R_k^2 have gamma distributions and, therefore,

$$f_{R_k}(r) = \frac{2(\rho\pi)^k r^{2k-1} e^{-\rho\pi r^2}}{(k-1)!}. \tag{6.5}$$

It follows that a simple way to generate a planar Poisson process is to construct a sequence of points with polar coordinates (R_k, Θ_k) $(k = 1, 2, \ldots)$, such that πR_1^2, $\pi(R_k^2 - R_{k-1}^2)$ $(k = 2, 3, \ldots)$ are independent and exponentially distributed with parameter ρ, while $\{\Theta_k\}$ is a sequence of independent variables each uniformly distributed on $[0, 2\pi)$ and independent of $\{R_k\}$.

(ii) *Doubly stochastic Poisson processes*

To generalize the doubly stochastic Poisson process of Section 3.3(iii) to a spatial process, we replace the unobserved rate process $\{\Lambda(t)\}$ by an unobserved process $\{\Lambda(\zeta)\}$ defined in the plane; a real-valued process in the plane is called a random field. If we restrict attention to second-order stationary processes, then

$$E\{\Lambda(\zeta)\} = \mu_\Lambda, \quad \operatorname{cov}\{\Lambda(\zeta + \tau), \Lambda(\zeta)\} = \gamma_\Lambda(\tau), \tag{6.6}$$

where γ_Λ is the autocovariance function and $\tau = (\tau_\xi, \tau_\eta)$.

For an arbitrary region B, we can calculate the mean and variance of $N(B)$ by first conditioning on Λ. We find that

$$E\{N(B)\} = \mu_\Lambda|B|, \quad \operatorname{var}\{N(B)\} = \mu_\Lambda|B| + \int_{B \times B} \gamma_\Lambda(\zeta - \zeta')d\zeta d\zeta'. \tag{6.7}$$

This compares with the general formula for an arbitrary orderly stationary point process of rate ρ,

$$\text{var}\,\{N(B)\} = \rho\,|B| + \int_{B \times B} \rho\,\{h(\zeta - \zeta') - \rho\}\,d\zeta\,d\zeta', \qquad (6.8)$$

where the conditional intensity function is defined by

$$h(\zeta) = \lim |\delta_\zeta|^{-1} \text{pr}\,\{N(\delta_\zeta) > 0\,|\,N(\delta_0) > 0\},$$

for a neighbourhood δ_0 of the origin and the limit is taken as the diameters of δ_ζ and δ_0 tend to zero. Thus the two-dimensional version of (3.42) is

$$h(\zeta) = \mu_\Lambda + \gamma_\Lambda(\zeta)/\mu_\Lambda. \qquad (6.9)$$

An important special case is of isotropic rate processes, for which

$$\gamma_\Lambda(\zeta) = \gamma_\Lambda^\dagger(r), \qquad (6.10)$$

say, where $r^2 = |\zeta|^2 = \xi^2 + \eta^2$. It follows that h also is isotropic; it can usefully be written as $h^\dagger(r)$. Given a point at $(0, 0)$, the probability of a point in the annulus $(r, r + \delta)$ is asymptotically $2\pi r h^\dagger(r)\delta + o(\delta)$. This function, or its three-dimensional generalization, arises, for example, in diffraction studies, under the name radial distribution function (Tabor, 1969, p. 196).

(iii) *Poisson cluster processes*
These are direct generalizations of one-dimensional Poisson cluster processes to spatial situations; in particular we consider the Neyman–Scott processes. We take as before an unobserved Poisson process of cluster centres with rate ρ_c, around each of which is distributed an independent subsidiary process or cluster. The cluster size is given by a variable M with probability generating function $G_M(z)$ and each point in the cluster is independently distributed about its cluster centre with probability density function f, where now f is a bivariate density function. The counting properties of the process are obtained as in the one-dimensional case. In particular, a spatial version of equation (3.51) for the conditional intensity function applies. If the density f is circularly symmetric it follows that h is isotropic. Because a cluster centre is not a point in the process, there is no loss of generality in taking the cluster centre as the centre of symmetry. A particular case is when f is bivariate normal with circular symmetry, so that

$$f(\xi', \eta') = \frac{1}{2\pi\sigma^2} \exp\left(-\frac{\xi'^2 + \eta'^2}{2\sigma^2}\right).$$

Then

$$h(\xi, \eta) = \rho_c E(M) + \frac{E\{M(M-1)\}}{E(M)} \frac{1}{4\pi\sigma^2} \exp\left(-\frac{\xi^2 + \eta^2}{4\sigma^2}\right).$$

(iv)   A process with independent locations

In Section 3.2(vi) a process is described which consists of a sequence of points with independent co-ordinates $\{T_k\}$ such that the distribution function of T_k is the k-fold convolution $G^{(k)}$ of G. An obvious generalization of this to a planar process, is to take a set of points independently located in $\mathbb{R}^2$ with independent rectangular co-ordinates $(\xi_{ij}, \eta_{ij})(i, j = 1, 2, \ldots)$, where Ξ_{ij} and H_{ij} have distribution functions $G_1^{(i)}$ and $G_2^{(j)}$, respectively. Like the one-dimensional process, this process is asymptotically Poisson far from the origin. If $N(t)$ is the number of points in the rectangle $(0, c_1 t] \times (0, c_2 t]$, then $E\{N(t)\} \sim O(t^2)$ while var $\{N(t)\} \sim O(t^{3/2})$ as $t \to \infty$, so that the index of dispersion is asymptotically zero as $t \to \infty$.

A further generalization of this process would be to suppose that Ξ_{ij} and H_{ij} are dependent variables with a constant correlation.

(v)   Renewal processes

We now illustrate briefly a situation in which direct generalization from one dimension does not achieve any very fruitful objective. A renewal process in one dimension can be defined by a sequence $\{X_1, X_2, \ldots\}$ of independent and identically distributed random variables representing intervals between successive points. In defining a renewal process in this way, the Xs are positive; for the purpose of the present argument it is enough that $E(X_i) = \mu_X > 0$. The conditional intensity function, or renewal density, $h(t)$ is such that $h(t) \to 1/\mu_X$ as $t \to \infty$, so that points are in a sense occurring at constant rate asymptotically.

A possible generalization to two dimensions is to replace each X_i by a vector, the vectors being independent and identically distributed with density $g(\zeta) = g(\xi, \eta)$. Then the conditional intensity at ζ is

$$\sum_{n=1}^{\infty} g^{(n)}(\zeta),$$

where $g^{(n)}$ denotes an n-fold convolution.

Within a region close to its mean, $g^{(n)}$ can be approximated by a bivariate normal density and the sum with respect to n by an integral. Further, to a first approximation, changes with n in the variances of the normal approximation can be ignored.

The details of the calculation can be simplified by a preliminary linear transformation to make the components of displacement uncorrelated and with unit variance, followed by a rotation of axes so that the ξ-axis is in the direction of mean motion. Thus the normal approximation to $g^{(n)}$ is

$$\frac{1}{2\pi n}\exp\left\{-\frac{(\xi-n\mu)^2}{2n}-\frac{\eta^2}{2n}\right\},$$

the approximation holding for (ξ,η) sufficiently close to $(n\mu,0)$. Suppose then that we wish to approximate the renewal density at the point $(v\mu,\sqrt{v}\eta')$, where v is a large integer and η' is fixed. We write $n=v+\sqrt{v}x$, and note that x is effectively bounded and moves in steps of $1/\sqrt{v}$, allowing us to write

$$\Sigma g^{(n)}(\xi,\eta)\sim\Sigma\frac{1}{2\pi n}\exp\left\{-\frac{(v\mu-n\mu)^2}{2n}-\frac{v\eta'^2}{2n}\right\}$$

$$\sim\Sigma\frac{1}{2\pi v}\exp\left(-\frac{\mu^2 x^2}{2}-\frac{\eta'^2}{2}\right)$$

$$\sim\int_{-\infty}^{\infty}\frac{1}{2\pi\sqrt{v}}\exp\left(-\frac{\mu^2 x^2}{2}-\frac{\eta'^2}{2}\right)dx$$

$$=\frac{1}{\sqrt{(2\pi v)}\mu}\exp\left(-\frac{\eta'^2}{2}\right);$$

recall that $v=\xi/\mu$, $\eta'=\eta/\sqrt{v}$.

The form of this relation could have been anticipated on general grounds. Considered in one dimension along the line of mean motion, the intensity tends to $1/\mu$ and in an orthogonal direction there is a diffusion of linear dimension $\sqrt{v}$. The process in no sense 'fills' the plane.

The result for an arbitrary co-ordinate system in which the components are possibly correlated and with unequal variances can be recovered by transformation, introducing the Jacobian, the square root of the determinant of the covariance matrix, to allow for scale change.

Other one-dimensional processes can be generalized similarly.

6.3 Some special constructions in two dimensions

In Section 6.2 we have seen that processes based on the Poisson process generalize straightforwardly from one to more than one dimension, but that for other processes an obvious generalization does not always yield a stationary two-dimensional process. We now consider briefly some processes which are not based on the Poisson process and which are genuinely multidimensional. These will be described in the plane although all can be defined equally simply in higher dimensional spaces.

(i)   Constructions using concentric circles

Suppose that two sequences of variables $\{A_i\}$ and $\{\Theta_i\}$ are used to construct a planar process in such a way that the points of the process have polar co-ordinates (R_i, Θ_i), where $\pi R_i^2 = A_1 + \ldots + A_i$ $(i = 1, 2, \ldots)$. Various assumptions can be made about the sequences $\{A_i\}$ and $\{\Theta_i\}$. For example, if $\{A_i\}$ and $\{\Theta_i\}$ are independent sequences, and the A_i are independent and identically distributed, then it can easily be shown that the process will have a constant rate only if the A_i are exponential variables and the Θ_i are marginally uniformly distributed on $[0, 2\pi)$. If, in addition, the Θ_i are independent then the process is a Poisson process, as mentioned in Section 6.2(i). Alternatively, if the A_i are independent exponential variables and the Θ_i are independent and identically, but not uniformly, distributed on $[0, 2\pi)$, then a non-stationary Poisson process is obtained.

Suppose that we take the A_i to be independent exponential variables with parameter ρ and the Θ_i to be dependent variables which are all marginally uniformly distributed. Then a simple form of dependency is for $\{\Theta_i\}$ to be a Markov sequence. To obtain second order properties of the constructed process, we need the conditional distribution of Θ_{i+j} given Θ_i for all i and j. One possibility for which this distribution is straightforward is to take Θ_1 to be uniformly distributed on $[0, 2\pi)$ and the conditional density of Θ_{i+j} given $\Theta_i = \theta_i$ to be the cardioid density

$$\frac{1}{2\pi}\{1 + 2\kappa^j \cos(\theta_{i+j} - \theta_i)\} \quad (i, j = 1, 2, \ldots; |\kappa| < 1/2). \quad (6.11)$$

It follows that Θ_i is marginally uniformly distributed for all i and the planar process has a constant rate ρ.

Second-order properties of the process can easily be derived. We can define a conditional intensity for the non-stationary process by

$$h(r', \theta'; r, \theta) = \lim |\delta_{r'\theta'}|^{-1} \mathrm{pr}\{N(\delta_{r'\theta'}) > 0 | N(\delta_{r\theta}) > 0\},$$

where we adapt the notation of Section 6.2 and let $\delta_{r'\theta'}$ denote a neighbourhood of the point with polar co-ordinates (r', θ') and take the limit as the diameters of $\delta_{r'\theta'}$ and $\delta_{r\theta}$ tend to zero. Thus, if $r' > r$

$$h(r', \theta'; r, \theta) = \sum_{j=1}^{\infty} \frac{1}{2\pi} \{1 + 2\kappa^j \cos(\theta' - \theta)\}$$

$$\times \frac{\rho\{\rho\pi(r'^2 - r^2)\}^{j-1}}{\Gamma(j)} e^{-\rho\pi(r'^2 - r^2)} 2\pi$$

$$= \rho[1 + 2\kappa \cos(\theta' - \theta) \exp\{-\rho\pi(1 - \kappa)(r'^2 - r^2)\}].$$

$$(6.12)$$

Then, for example, the variance of $N(B)$ for any bounded set B is given by

$$\text{var}\{N(B)\}$$

$$= \rho|B| + 2\rho^2 \int_{\{(r, \theta) \in B\}} r dr d\theta \int_{\{(r', \theta') \in B : r' \geq r\}} r' dr' d\theta' 2\kappa \cos(\theta' - \theta)$$

$$\times \exp\{-\rho\pi(1 - \kappa)(r'^2 - r^2)\}.$$

$$(6.13)$$

If B is a sector of an annulus centred on the origin it follows from (6.13) that $N(B)$ is overdispersed or underdispersed relative to a Poisson variable according to whether κ is positive or negative.

The construction described above places one point randomly on each of a set of concentric circles. More generally we can locate a random number of points on each circle. One very simple possibility locates the points on each circle independently as follows. Let $\{A_i\}$ be a sequence of independent exponentially distributed variables with parameter ρ and define R_i by $\pi R_i^2 = A_1 + \ldots + A_i$ as before. Let $\{M_i\}$ be a sequence of independent non-negative integer valued variables all with probability generating function $Q(z) = \Sigma q_n z^n$ such that $E(M_i) = \Sigma n q_n < \infty$. Given $M_i = m_i$ $(i = 1, 2, \ldots)$, let $\{\Theta_{ij}; j = 1, \ldots, m_i; i = 1, 2, \ldots\}$ be a sequence of independent and identically distributed variables on $[0, 2\pi]$. Now define a planar process in which the points have polar co-ordinates $\{(R_i, \Theta_{ij}); j = 1, \ldots, m_i; i = 1, 2, \ldots\}$. Note that we may take $q_0 = 0$ without loss of generality. It can be shown that this process is overdispersed with respect to a Poisson process as long as $E\{M_i(M_i - 1)\} > 0$. Since $q_0 = 0$, $E\{M_i(M_i - 1)\} = 0$ if and only if M_i is one with probability one, and we already know that in this case the process is a non-stationary Poisson process.

If we look at the process far from the origin then the concentric circles will be arbitrarily close. However, for a suitably well-behaved distribution of Θ_{ij}, the probability that a point on a particular circle falls in some fixed bounded set will be small and the probability of two such points will be negligible. Thus it is intuitively reasonable that the process, far from the origin, will closely resemble a Poisson process, and this can be proved without difficulty (Isham, 1977a).

Finally, we note that, if $E(M_i)$ is at all large, the inbuilt regularity of this process is likely to be noticeable. One way of removing some of the obvious regularity is to displace the points independently, possibly along lines through the origin.

(ii) A Markov construction

In one of the processes discussed above, a simple form of dependency between the points was introduced via a Markov sequence. Another way of using a Markov sequence is as follows. Again it is convenient to work in polar co-ordinates. Let $\{Z_i\}$ be a stationary Markov sequence of points, where Z_i has polar co-ordinates (R_i, Θ_i). Then define a sequence $\{P_n\}$ of finite point processes such that P_n consists of the n points $(\sqrt{nR_i},\ \Theta_i)$ $(i = 1, \ldots, n)$. Thus $\{P_n;\ n = 1, 2, \ldots\}$ is a spatial point process evolving in discrete time, where at each time instant one further point is added to the existing ones and the radial co-ordinates of all of them are rescaled. The scaling ensures that the average density of points of P_n does not depend on n.

The properties of the process P_n for any fixed n are straightforward to write down in terms of the transition function $F(.\,;\zeta)$ for the Markov sequence, that is the conditional distribution of Z_{i+1} given $Z_i = \zeta$ $(i = 1, 2, \ldots)$, and of the marginal distribution of Z_i. An example is to take the radial co-ordinates R_i of the Markov sequence to be independent and such that πR_i^2 is uniformly distributed on $[0, 1/\rho)$, while the angular co-ordinates Θ_i are a dependent sequence of cardioid variables with conditional density given by (6.11), which are independent of $\{R_i\}$. As before Θ_1 is uniformly distributed over $[0, 2\pi)$. The main difference between this construction of a finite process P_n and the similar construction in Section 6.3(i), is that there is the ordering of the angular variables given by the Markov sequence is the same as the ordering of the points in increasing distance from the origin, while here there is no such correspondence.

If the Markov sequence $\{Z_i\}$ is in some sense well-behaved, then it is intuitively reasonable that P_n closely resembles a Poisson process when n is sufficiently large. For, if a particular point of P_n lies in some fixed set B, then the corresponding member of $\{Z_i;\ i = 1, \ldots, n\}$ must be in the set $\{(r, \theta):(\sqrt{nr}, \theta) \in B\}$, which is in the neighbourhood of the origin and has an area which is $O(1/n)$. If the probability that Z_i is in this set is small and also the dependence between Z_i and Z_{i+j} is asymptotically negligible as $j \to \infty$ for fixed i, then P_n consists of a large number of points each of which is 'almost independent' of most of the rest and which has a very small probability of falling in B. One may then appeal to a limit theorem for a superposition of point processes, as described in Section 4.5, to infer that P_n must be approximately a Poisson process. Sufficient conditions for this result to hold are discussed by Isham (1977b).

(iii)   *Lattice-based processes*

Various planar processes can be constructed using a lattice. For example one may take the direct product of two one-dimensional processes. If the points of the one-dimensional processes have co-ordinates $\{\xi_i\}$ and $\{\eta_j\}$, respectively, then their direct product has points with rectangular co-ordinates (ξ_i, η_j) $(i, j = 1, 2, \ldots)$. Thus the direct product gives a 'random lattice', the properties of which can be obtained from those of the component processes. This process seems unrealistic physically; some more natural alternatives are based on a regular lattice.

One possibility is to perturb each of the lattice points, according to some distribution; a way of regarding the resulting point process is, therefore, as a cluster process in which the cluster centres are the lattice points and there is one point in each cluster, while another way is to view the perturbed lattice as a translation of the regular lattice. The operation of translation on point processes was discussed in Section 4.4, but since here the original process on which the operation is performed is deterministic, it is probably preferable to derive any required properties directly.

Another possibility is to suppose that only some of the points in the regular lattice are occupied by points of the process. That is, a binary variable is defined at each lattice point indicating presence or absence of a process point. Such processes have been extensively studied in various contexts, including that of spin systems in physics, where the binary variable indicates direction of spin. Bartlett (1975) and Besag (1974) have discussed such lattice processes, with some emphasis on statistical analysis; their discussion is, however, not restricted to binary variables.

Both the above types of regular lattice model have been considered in connection with the positions of plants. In the first kind of process it is assumed that an original intention to space plants regularly is distorted by random positioning of the plants around the lattice points, whereas in the second type of process it is assumed that the plants are originally regularly spaced but that some die.

6.4 Gibbs processes

In some contexts, it is natural to suppose that the configuration of points is influenced by interactions between points, in particular by processes of attraction and repulsion between points, or incipient points, relatively close together. A fairly general approach to such possibilities is as follows.

Consider an orderly stationary process, say in some Euclidean space, and examine the positions of the points in a bounded region B. For a Poisson process of rate ρ, the probability density for n points at $\zeta_1, \ldots, \zeta_n$ is

$$\rho^n e^{-\rho|B|}, \tag{6.14}$$

where the probability density specifies in the usual way the probability of finding n points in small neighbourhoods of $\zeta_1, \ldots, \zeta_n$, there being no other points in B. Note that to avoid multiple counting, we regard (6.14) as defined with some arbitrary ordering of position. Alternatively, we can suppose the points to be numbered arbitrarily $1, \ldots, n$ and then take

$$\rho^n e^{-\rho|B|}/n! \tag{6.15}$$

as giving the probability density that the first point is at ζ_1, the second at ζ_2 and so on. Then (6.15) is defined over all $\zeta_i \in B(i = 1, \ldots, n)$, i.e. over the set $B^n = B \times \ldots \times B$. The results of Section 3.1(i) follow immediately from (6.15), namely that the number, $N(B)$, of points in B has a Poisson distribution of mean $\rho|B|$ and that given $N(B) = n$, the points are independently and uniformly distributed over B.

Now suppose that for the process we are considering there are, for each n, functions $g_n(\zeta_1, \ldots, \zeta_n)$ which are symmetrical in their arguments and such that the probability density for the event that there are exactly n points in B with the first point at ζ_1, the second point at $\zeta_2, \ldots$ is

$$cg_n(\zeta_1, \ldots, \zeta_n)\rho^n e^{-\rho|B|}/n!, \tag{6.16}$$

where c is a normalizing constant, determined by integrating (6.16) with respect to $\zeta_1, \ldots, \zeta_n$ and summing over n. Note that

$$\text{pr}\{N(B) = n\} = \frac{c\rho^n e^{-\rho|B|}}{n!} \int_{B^n} g_n(\zeta_1, \ldots, \zeta_n)d\zeta_1 \ldots d\zeta_n \tag{6.17}$$

and hence that, given $N(B) = n$, the points are distributed over B with a density proportional to g_n. It would, of course, be possible to absorb the normalizing constant c into the functions g_n.

Thus, by (6.16), g_n specifies how likely a particular set of points is relatively to a Poisson process of rate ρ. One interpretation, if, for all n, $g_n(.) \leq g_{max}$, is that we generate points from a Poisson process of rate ρ and 'accept' the result with probability $g_n(\zeta_1, \ldots, \zeta_n)/g_{max}$ and otherwise 'reject' the result. We continue until a set is accepted. Mathematically cg_n specifies the density of the probability measure of the point process relative to that of the Poisson process of rate ρ, i.e. is a so-called Radon–Nikodym derivative of the first measure with respect to the second.

We can always write

$$g_n(\zeta_1, \ldots, \zeta_n) = \exp\{-\phi_n(\zeta_1, \ldots, \zeta_n)\}, \tag{6.18}$$

provided that we allow formally infinite ϕ_n. If we do this and consider ϕ_n as a potential function for the process, we call the process a Gibbs process. Given $N(B) = n$, the conditional density of positions is

$$\exp\{-\phi_n(\zeta_1, \ldots, \zeta_n)\}/\int_{B^n} \exp\{-\phi_n(\zeta_1', \ldots, \zeta_n')\} d\zeta_1' \ldots d\zeta_n'. \tag{6.19}$$

The use of Gibbs's name comes from the connection with classical statistical mechanics. Here the state of a system with m molecules is specified by one point in the $6m$ dimensional phase space of positions and momenta: let $\phi(\zeta)$ denote a potential function depending on the position co-ordinates. Then the so-called Gibbs canonical ensemble for a system embedded in, exchanging energy with, and in thermal (i.e. statistical) equilibrium with a much larger similar system, specifies a probability density in the 'position' part of the phase space proportional to $\exp\{-l\phi(\zeta)\}$, where the constant l is interpreted as $(kT)^{-1}$, k being Boltzmann's constant and T being absolute temperature; see, for example, Kittel (1958, Section 11). Separation of the position and momentum contributions is possible if the total energy is the sum of a potential energy depending only on the positions and a kinetic energy depending only on the momenta.

So far, (6.18) has put no major restriction on the point process. If, however, we require that all the ϕ_n are of essentially the same structure, special families of point processes are produced. One restriction is to suppose that

$$\phi_n(\zeta_1, \ldots, \zeta_n) = n\psi_1 + \sum_{i_1 > i_2} \psi_2(\zeta_{i_1} - \zeta_{i_2}) + \sum_{i_1 > i_2 > i_3} \psi_3(\zeta_{i_1} - \zeta_{i_2}, \zeta_{i_1} - \zeta_{i_3})$$

$$+ \ldots + \sum_{i_1 > \ldots > i_s} \psi_s(\zeta_{i_1} - \zeta_{i_2}, \ldots, \zeta_{i_1} - \zeta_{i_s}). \tag{6.20}$$

Thus $\psi_2(\zeta) = \psi_2(-\zeta)$ is the contribution to the potential from a pair of points whose vector separation is ζ. The ψ_j are even functions of each of their arguments. An interesting case is $s = 2$, when the potential is formed by summing contributions over all possible pairs of points and can be written

$$\phi_n(\zeta_1, \ldots, \zeta_n) = n\psi_1 + \sum_{i>j} \psi(\zeta_i - \zeta_j). \qquad (6.21)$$

The following comments can be made about (6.20) and (6.21). First it is clear from (6.16) and (6.20) that the density depends on the parameter ρ of the reference Poisson process only via the combination $\rho e^{-\psi_1}$, so that changes in ρ can be accommodated in changes in the 'self potential' ψ_1. Next, restrictions are needed to ensure that (6.17) defines a proper probability distribution, i.e. to prevent explosion to an infinite number of points. For example, if in (6.21), for all arguments ζ, $\psi(\zeta) < -\delta < 0$, then there will be a factor in (6.17) of at least $\exp\{\frac{1}{2}n(n-1)\delta\}$, which for large n dominates other contributions, and makes convergence to a proper distribution for $N(B)$ impossible. Finally, a careful treatment requires attention to boundary conditions. Physically, the most satisfactory procedure is to imagine the reference region B imbedded within a very much larger region, throughout which the process (6.20) or (6.21) holds, and then to examine the process within B, integrating out over the process outside B. This is entirely analogous to the procedure mentioned briefly above in connection with the canonical ensemble of statistical mechanics. In a general way it is clear that provided that B is large and that the functions $\psi_2, \ldots, \psi_s$ in (6.20) are appreciable only for points close together, the behaviour outside B affects only points in B near the boundary and hence precise specification of boundary conditions is not crucial.

In studying special cases of (6.20) and (6.21) it is convenient either to put $\psi_1 = 0$, modifying ρ appropriately, or to put $\rho = 1$, modifying ψ_1. One special case of (6.21) is when there is a critical radius r_0 such that particles interact only if less than r_0 apart, i.e. $\psi(\zeta) = 0$ if $|\zeta| > r_0$. An extreme special case is the 'hard core' model in which $\psi(\zeta) = \infty$ ($|\zeta| \leq r_0$), when we have a Poisson process of rate ρ in which only realizations in which no two points are nearer than r_0 are acceptable. Sinai (1963) has proved the ergodicity of the hard core model. Another special case is $\psi(\zeta) = \psi'(|\zeta| \leq r_0)$, $\psi(\zeta) = 0(|\zeta| > r_0)$).

It is natural to ask whether Gibbs processes have any special properties other than the relation to the Poisson process implied in the definition. Provided that all the dependencies in the potential are short range we can argue as follows. Consider the process in a region $C \subset B$. We can divide the whole region B into (i) a suitably defined boundary layer consisting of elements of B which are not in C but which are close to the frontier of C, (ii) C itself, now to be called the inside of C, and (iii) the remainder of B remote from C. Given the points in the boundary layer, the potential function for the process inside C, region (ii), does not involve the positions of points in region (iii), and vice versa. That is, given the process in the boundary layer, the processes in the remaining parts are conditionally independent. This is a generalization of the Markov property of temporal processes which can be phrased in the form that, given the process at time t, the 'future' of the process after t is independent of the 'past' of the process before t. Put alternatively, the conditional density in C given the process in the rest of B depends only on the process in the boundary layer. The application of this idea to modern constructive quantum field theory is discussed by Simon (1974, p. 94).

The above argument is very general but a bit vague. A precise formulation requires stronger restrictions on the potential function and is briefly as follows. Suppose that some pairs of elements of B are called neighbours: if ζ_1 is a neighbour of ζ_2, then ζ_2 is a neighbour of ζ_1, and each element is its own neighbour. The simplest illustration is to call two elements neighbours if and only if they are less than r_0 apart. Call a collection of elements a clique if any two elements in the collection are neighbours. Suppose now that in the form (6.20), the only non-zero ψ_is arise when the ζs involved form a clique. This is one precise formulation of the earlier vague idea of short-range interactions.

The resulting point process has a Markov property in the following precise sense. Consider as before a region $C \subset B$. Let the boundary layer of C, denoted by $\mathscr{L}(C)$, be the collection of elements in B but not in C which are neighbours of at least one element of C. Suppose now that we compute the conditional density for positions in C, given all the points outside C. Then the conditional density of n points at $\zeta_1, \ldots, \zeta_n$ in C, given m points at $\zeta'_1, \ldots, \zeta'_m$ in $\mathscr{L}(C)$ and l points at $\zeta''_1, \ldots, \zeta''_l$ in the remainder of B is, from (6.16) and (6.18), proportional to

$$\frac{\exp\{-\phi_{n+m+l}(\zeta_1,\ldots,\zeta_n,\zeta'_1,\ldots,\zeta'_m,\zeta''_1\ldots,\zeta''_l)\}}{\sum_p (p!)^{-1} \int_{C^p} \exp\{-\phi_{p+m+l}(\zeta'''_1,\ldots,\zeta'''_p,\zeta'_1,\ldots,\zeta'_m,\zeta''_1,\ldots,\zeta''_l)\}}$$

$$\times \, d\zeta'''_1 \ldots d\zeta'''_p, \quad (6.22)$$

where, as noted previously, we can take $\rho = 1$ without loss of generality.

Now in the form (6.20), there are no non-zero terms involving a ζ

and a ζ'', and the terms in (6.22) involving a ζ' and a ζ'' are the same in numerator and denominator and hence cancel, as do the terms purely in ζ's and purely in ζ''s. Hence (6.22) does not depend on $\zeta''_1, \ldots, \zeta''_l$, i.e. on the number and position of points in the part of B outside $C \cup \mathscr{L}(C)$. It can be shown that the only point processes with this kind of Markov structure are those with potentials of the form (6.20) in which the summations are restricted to cliques: the main further restriction required is that if some collection of points has positive density so does any subset. For details and for a lucid formulation giving careful attention to mathematical proprieties, see Ripley and Kelly (1977).

It throws some light on the relation between Gibbs processes and the Markov property discussed above to consider a one-dimensional renewal process. This has a Markov property in the quite simple form that given that there is a point at spatial position ξ, the process to the left of ξ is independent of the process to the right of ξ. Nevertheless, this is not the precise form of Markov property involved in the discussion centring on (6.22). For there, in order to define the boundary layer, we had to define a neighbour relation between elements of the space rather than a relation between points of the process. If the intervals between successive points in the renewal process are with probability one less than some constant a, we could define two elements to be neighbours if they are less than a apart. The boundary layer of an interval $C = [c_1, c_2]$, would then be the union of $(c_1 - a, c_1)$ and $(c_2, c_2 + a)$. There will always be at least one point in each of these intervals, and it is clear that given the points in the boundary layer, the processes in C and the remainder of B are independent, a correct but much weakened form of the Markov property based on a point of the process. Even this weakened property fails if the intervals in the renewal process are unbounded.

It is normally very difficult to obtain explicitly properties of processes defined in the Gibbs form. For the techniques used by theoretical physicists to study such systems in connection with imperfect gases and dense fluids, see, for example, Croxton (1974, Chapters 1 and 2).

6.5 Spatial-temporal processes

(i) *Introduction*
We now consider spatial processes followed through time. While again the most important case is probably two-dimensional space, the arguments are general. For graphical representation it is convenient

to take one-dimensional space and to represent the process in terms of co-ordinates (t, ζ) with respect to a horizontal time axis and a vertical space axis.

There are several different types of spatial–temporal process. First there is a broad distinction between processes of points in space–time and processes in which a point in space has an existence through time, at least for a limited period. Fig. 6.3(i) illustrates a point process in space–time and Fig. 6.3(ii)-(iv) show various processes of the second

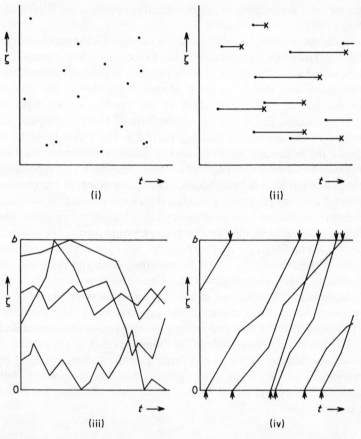

Fig. 6.3. *Spatial–temporal point processes. t, time co-ordinate. ζ, spatial co-ordinate. (i) Points in space-time, (ii) ., birth; ×, death. Points have fixed position in space. Spatial point process at any time. (iii) Fixed number of individuals moving between reflecting barriers. Spatial point process at any time. (iv) ↑, entry at 0. ↓, exit at b. Spatial point process at any time. Temporal point process at any spatial level.*

kind. An example of the first kind is provided by the instant and place of identification of a case of an infectious disease. Another is the recording of cosmic ray showers at ground level.

For processes in which 'points' continue for a while in time, we may have to consider controlling processes of birth and death, i.e. the creation of new points and the extinction of points occurring within the spatial region B of interest; of movement of points within B; and of immigration and emigration, i.e. the entry and exit of points across the boundary of B. In particular, the following three special cases are important. First suppose that there is no movement, immigration or emigration; see Fig. 6.3(ii). Points are born and keep a fixed position in space until they die. Such processes are relevant in botany, for instance. Secondly, there may be movement, but no birth, death, immigration or emigration. That is, a fixed number of points perform random movements within B. Mathematically, this case may arise if B is closed, for example if B is the circumference of a circle, or a torus, or if the boundary of B is a reflecting barrier. Physically, the case is more likely to arise when B is large and fairly short time periods are considered, so that immigration and emigration, while present, can be neglected as a first approximation. A third special case occurs when there is no birth or death, points appearing by immigration across the boundary of B. They move within B until emigrating across the boundary of B; see Fig. 6.3(iv). An example in traffic studies is of cars entering, traversing and then leaving a section of road.

This second category of process, illustrated in Fig. 6.3(ii)–(iv), is such that a cross-section of the process at a particular time forms a spatial point process. Such processes may be studied either because the progression of the spatial process in time is of intrinsic interest, or because it is hoped that the equilibrium distribution in time of such a process will provide a reasonable model for, and explanation of, a spatial process observed at a single time point.

We are able to give only a preliminary sketch of the theory of spatial-temporal processes.

(ii) *Processes of points in space–time*

We now discuss briefly point processes in space–time. Clearly the simplest such process is the Poisson process; the rate ρ has dimension [area × time]$^{-1}$. An important generalization is to allow heterogeneity in either or both of space and time. For instance, in the epidemiological example mentioned above, heterogeneity in space might correspond largely to variation in population density and variation in time to seasonal variation. The simplest form of process would then be a non-homogeneous Poisson process with rate

$$\rho(t, \zeta) = \rho_{(1)}(t)\rho_{(2)}(\zeta). \tag{6.23}$$

It would also be possible for space- and time-components of the rate to combine in some other simple way, for example additively.

An important qualitative concept in the study of point processes in space-time is of space × time interaction. Such a concept needs careful definition, at least if it is to be given physical interpretation. In a stationary process, however, a positive such interaction means a relatively high value for the second-order intensity function $h(t, \zeta)$, for small $t > 0$ and small ζ. That is, a point is relatively likely to be followed soon by another point close by. Clearly several of the models of Chapter 3 might be relevant, in particular the Neyman–Scott cluster process and the doubly stochastic Poisson process. If the process is not stationary, the model (6.23) is a natural reference point. An alternative would be to aim for a transformation to operational time and to new spatial co-ordinates to induce stationarity. While the non-stationary case is important in statistical applications we shall not consider it further.

(iii) *Non-interacting Poisson processes with birth, death and movement*
We can outline a fairly general treatment of processes in which new points are born in accordance with a Poisson process of rate ρ, say, where life-times are independently and identically distributed with density g and which move in accordance with some random process realized independently for each point. The essential requirement for a simple discussion is the absence of interactions between different points.

In studying the joint distribution of the numbers of points in two spatial regions C_1 and C_2 at times t_1 and t_2, $N(C_1; t_1)$, $N(C_2; t_2)$, say, the bivariate Poisson distribution plays a central role: see Section 5.3(i) and Exercise 1.8.

It is convenient to decompose the process by first considering a discrete distribution of life-time with values $\{l_i\}$ having probabilities $\{p_i\}$. An arbitrary distribution can always be produced by a limiting operation. Then we consider individuals of constant life-time l_i produced in a Poisson process of rate ρp_i, the processes for different i being independent. We have considered such a decomposition in virtually the present context in Section 5.6(iii), so that in the following discussion some details are omitted.

Suppose first that there is no movement; see Fig. 6.3(ii). Then for a fixed spatial set C considered at time t, a point of life-time l_i occurs in

C if and only if it is born in a region of total 'volume' $l_i|C|$. Hence the number of such points has a Poisson distribution of mean $\rho p_i l_i |C|$. By the additive property of the Poisson distribution, the total number of points in C at time t thus has a Poisson distribution of mean

$$\sum \rho p_i l_i |C| = \rho E(L)|C|; \tag{6.24}$$

compare this with the result of Section 5.6(iii), where the factor $|C|$ does not arise.

An extension of this argument involving a collection of sets C_1, $C_2, \ldots$ and the defining properties of the Poisson process shows easily that the spatial process at time t is a Poisson process of rate $\rho E(L)$.

Consider now the same spatial set C at two different times t_1 and $t_1 + t$. The random variables $N(C; t_1)$ and $N(C; t_1 + t)$ have a bivariate Poisson distribution and it is necessary to find only the covariance, the expected number of common points.

The argument leading to (5.39) shows easily that the required covariance is

$$\rho|C| \int_t^\infty (l - t)g(l) \, dl = \rho|C| \int_t^\infty \mathscr{G}(l) \, dl, \tag{6.25}$$

where $\mathscr{G}$ is the survivor function of life-time. If, in particular, g is an exponential distribution, $g(x) = \mu_L^{-1} e^{-x/\mu_L}$, with $\mu_L = E(L)$, then

$$\operatorname{cov}\{N(C; t), N(C; 0)\} = \rho|C|\mu_L e^{-t/\mu_L}, \tag{6.26}$$

in line with the Markovian character of the process as it evolves in time.

These results establish clearly the simple structure of the process when there is no movement. In this and in more general cases with movement, the joint distribution of counts in spatial sets at k time points is a k-variate Poisson distribution.

Suppose now that movement occurs. We require the process of movement to be independent from point to point, and to be independent of the time and spatial position of birth. The probabilistic properties of the movement may, however, depend on life-time; or, equivalently, life-time may depend on the motion of the point in space, although not on absolute position in space. Then by considering first points with a fixed life-time l_i and decomposing the movements into a discrete set of possibilities, it follows again that at any time the spatial process is a Poisson process of rate $\rho E(L)$.

One description of the dependence in time is derived from the bivariate Poisson distribution of $N(C_1; t_1)$ and $N(C_2; t_2)$, again characterized by the covariance, or equivalently the expected number of common points. As before set $t_1 = 0$ and $t_2 = t > 0$. Take as spatial origin some convenient reference point in C_1, and denote by

$$P_{C_1, C_2}(t; -u, \zeta, l), \tag{6.27}$$

the probability that a point given to have life-time l and born at position ζ at time $-u$ is in C_1 at time 0 and in C_2 at time t. Consider first only the points of life-time l_i, born in a Poisson process of rate ρp_i. Then the expected number of common points is

$$\rho p_i \int_B d\zeta \int_0^{\max(l_i - t, 0)} du\, P_{C_1, C_2}(t; -u, \zeta, l_i), \tag{6.28}$$

so that the required covariance is, in general,

$$\rho \int_B d\zeta \int_t^\infty dl\, g(l) \int_0^{l-t} du\, P_{C_1, C_2}(t; -u, \zeta, l). \tag{6.29}$$

In principle, this can be found once the hitting probabilities (6.27) are evaluated.

A special case arises when the points perform Brownian motion, possibly with drift, when (6.27) can be evaluated via multivariate normal integrals. We consider for simplicity one-dimensional space and take, without essential loss of generality, C_1 to be the interval $(0, 1)$ and C_2 to be the interval (a, b). Because of the special properties of Brownian motion, we can argue as follows. The expected number of points of life-time l_i that are in C_1 at $t = 0$ and are still alive at time t is $\rho p_i \max(l_i - t, 0)$. Given such a point is in C_1, its position, ξ, is uniformly distributed in $(0, 1)$. Further, given ξ, the probability that the point is in C_2 at time t is

$$\Phi\left\{\frac{b - \mu(l_i)t - \xi}{\sigma(l_i)\sqrt{t}}\right\} - \Phi\left\{\frac{a - \mu(l_i)t - \xi}{\sigma(l_i)\sqrt{t}}\right\} = \psi(t, l_i, \xi),$$

say, where $\mu(l)$ and $\sigma^2(l)$ are the drift and diffusion parameters for points of life-time l. The contribution to the covariance is thus

$$\rho p_i \max(l_i - t, 0) \int_0^1 \psi(t, l_i, \xi)\, d\xi$$

$$= \rho p_i \max(l_i - t, 0)\Psi(t, l_i),$$

where Ψ is easily expressed in terms of the standard normal distribution function and density.

It follows that in the general case, the required covariance is

$$\rho \int_t^\infty (l - t)\Psi(t, l)g(l)\, dl. \tag{6.30}$$

Note that if the parameters of the Brownian motion are independent of l, comparison of (6.30) with (6.25), with $|C| = 1$, shows that the function $\Psi(t)$ gives the deflation of covariance arising from the random movement.

We repeat that the essentially very simple structure of the process is a consequence of the absence of interaction between points.

(iv) *Processes with immigration*

As a final instance of a process without interaction, consider, in one spatial dimension, points immigrating across a boundary at, say, $\xi = 0$ in a stationary orderly point process in time. Each point then pursues an independent path across the region of interest, before emigrating across the boundary at $\xi = b$, say. Fig. 6.3(iv) illustrates the situation. For some purposes it may be sensible to ignore emigration, i.e. to let $b \to \infty$. This process provides an approximate representation of a stream of traffic entering a long section of wide road, e.g. a motorway, through a constriction. The constriction imposes an input process that, locally at least, may be underdispersed relative to the Poisson process. It is assumed that once through the constriction the cars are free to move independently, a special case being where each car has constant speed, the speeds of different cars being independent and identically distributed random variables.

Various properties of the process may be of interest. For example we may examine the temporal point process at the boundary b. Let the random variable T represent the time during which a point is in $[0, b]$ between entry and emigration. We assume that the variables T corresponding to different points are independent and identically distributed. Then the temporal process at b is a random translation of the process at 0, as described in Section 4.4. From the results of that section we know that if the distribution of T is sufficiently dispersed relative to the mean interval between input points, then the output process at b is approximately a Poisson process. At the other extreme, if T is a constant with probability one, which corresponds to the points moving with a constant velocity across the region, the output process is exactly the input process, except for a shift in origin.

We can also consider the properties of the spatial process at a fixed time. If all the points move from 0 to b, with the same constant speed s, then the spatial process in $[0, b]$ at time t is simply a rescaled version of the input process for the time interval $[t - b/s, t]$. The distances between points are the corresponding time intervals scaled by a factor s. More generally, suppose that points immigrate across the $\xi = 0$ boundary at times $\ldots t_{-1} < t_0 \leq 0 < t_1 < t_2 \ldots$ and let $p(\xi; t)$

be the probability density of the position of a point a time t after immigration. Then, for example, the number of points in $A \subset [0, b]$ at time t can be represented by

$$\sum_{\{n:\, t_n \leq t\}} \chi_n(A),$$

where $\chi_n(A)$ is an indicator variable, taking value one if the nth process point is in A at time t, i.e.

$$\operatorname{pr}\{\chi_n(A) = 1\} = \int_A p(\xi; t - t_n) d\xi.$$

In particular, if the density of the time taken to reach A from the origin is sufficiently flat then the spatial process of points in A at a fixed time will be close to a Poisson process.

(v) *Systems with interaction*

Processes in which birth, death and movement involve interactions between points are much more difficult to study. Even the process with birth and death but no movement, but in which birth and death rates depend on other points in the neighbourhood in question, raises difficult issues and no comprehensive treatment is available at the time of writing. If we concentrate on the spatial process, taking the equilibrium distribution in time, some general results are available and these are now outlined. There is a strong link with classical statistical mechanics, the objective of which is to find equilibrium properties from a Hamiltonian function without explicit consideration of the dynamics of the process.

Consider first processes with birth and death but no movement. It is assumed that the birth and death rates at time $t +$ are determined by the spatial process at t. This rules out the possibility allowed in Section 6.5(iii) that the probability of death depends on the length of time since the particular point was born.

Suppose that there, are at time t, points at $\zeta_1, \zeta_2, \ldots$ in the large reference region B. Assume that the probability of a birth in time $(t, t + \delta_t)$ in a small region of volume $|\delta_\zeta|$ centred on spatial point ζ is

$$b(\zeta; \zeta_1, \zeta_2, \ldots) |\delta_\zeta| \delta_t + o(|\delta_\zeta|) o(\delta_t).$$

Further if there is a point at ζ and other points at $\zeta_1, \zeta_2, \ldots$, assume that the probability of death in $(t, t + \delta_t)$ of the individual at ζ, is

$$d(\zeta; \zeta_1, \zeta_2, \ldots) \delta_t + o(\delta_t).$$

For spatial stationarity we require b and d to be functions of the vector displacements $\zeta - \zeta_1, \zeta - \zeta_2, \ldots$ and for temporal stationarity b and d must, as the notation implies, be independent of t.

In principle, we may now proceed as follows. Make the problem discrete both in space and time. For the former divide B into a large number, n_c, of cells of equal volume and such that with very high probability each cell contains either one point or no point. Then the process can be considered as a discrete-state discrete-time Markov chain, and its properties, and in particular its equilibrium distribution, can be determined from the transition matrix of the chain. Passage to the limit as $n_c \to \infty$ and with the time interval tending to zero gives the properties of the original system. Of course there are major problems in making rigorous the conclusions from the limiting operation, but much more serious is that the number of states in the Markov chain, 2^{n_c}, is very large even for modest n_c, which makes explicit computation difficult.

To make further progress we restrict attention to reversible processes. For a discrete-state discrete-time Markov chain with transition matrix $((p_{ij}))$ and equilibrium distribution $\{\tilde{\omega}_i\}$, reversibility requires that transitions $i \to j$ occur as often as transitions $j \to i$, so that

$$\tilde{\omega}_i p_{ij} = \tilde{\omega}_j p_{ji}. \tag{6.31}$$

This is called the equation of detailed balance, and is a much stronger requirement than that of statistical equilibrium, under which transitions into and out of state i, regardless of the type of transition, balance. In our application we may take $i \to j$ to correspond to birth of a point near ζ, while $j \to i$ corresponds to death of a point near ζ. Then if, in addition to any point at ζ, the state contains points at $\zeta_1, \ldots, \zeta_n$, we have in an obvious notation

$$p_n(\zeta_1, \ldots, \zeta_n) b(\zeta; \zeta_1, \ldots, \zeta_n) = p_{n+1}(\zeta, \zeta_1, \ldots, \zeta_n) d(\zeta; \zeta_1, \ldots, \zeta_n). \tag{6.32}$$

We now return to the continuous version of the problem using the notation of Section 6.4 in terms of the Gibbs distribution for p_n, and taking a Poisson process of unit rate as the reference standard. Then (6.16), (6.18) and (6.32) give

$$\frac{b(\zeta; \zeta_1, \ldots, \zeta_n)}{d(\zeta; \zeta_1, \ldots, \zeta_n)} = \exp\{-\phi_{n+1}(\zeta, \zeta_1, \ldots, \zeta_n) + \phi_n(\zeta_1, \ldots, \zeta_n)\}. \tag{6.33}$$

This simplifies if the representation (6.20) involving interactions of finite numbers of points is used and if, in particular, (6.21) with two-point interactions is used,

$$\frac{b(\zeta; \zeta_1, \ldots, \zeta_n)}{d(\zeta; \zeta_1, \ldots, \zeta_n)} = \exp\left\{-\psi_1 - \sum_{i=1}^{n} \psi(\zeta - \zeta_i)\right\}. \tag{6.34}$$

It follows that if, given b and d, there is a process with a unique equilibrium distribution and if (6.33) is satisfied for some $\{\phi_n\}$, then the process is reversible and the equilibrium distribution is determined. Conversely, given an equilibrium distribution we may be able to find b and d to generate it: indeed because only the ratio b/d is relevant there will typically be many possibilities, including situations in which either b is constant or d is constant.

This discussion raises many problems, some of which are as follows:

what stationary spatial processes cannot be recovered as equilibrium distributions for reversible birth–death processes without movement;

what extra generality is obtained by allowing the death rate to depend on the age of the point and, in particular, does this generality show only in the transient properties of the process, i.e. in correlations through time;

what are the properties of the above processes in time and in particular does the reversibility exclude any kind of quasiperiodic behaviour (Whittle, 1975)?

We now consider processes with movement but no birth or death. In line with the above discussion, we assume that movement is a Markov process determined by the current spatial distribution. We concentrate on movements which are Brownian motion with a smoothly varying vector drift parameter $\mu(\zeta; \zeta_1, \ldots, \zeta_n)$ for a point at ζ, given other points at $(\zeta_1, \ldots, \zeta_n)$. That is, the expected movement in time δ_t is $\mu(\zeta; \zeta_1, \ldots, \zeta_n)\delta_t + o(\delta_t)$. We suppose also that, possibly after a preliminary transformation of spatial co-ordinates, the components of displacement are independent with variance $\sigma^2 \delta_t + o(\delta_t)$. We pass to a discrete version of the problem and apply the equation of detailed balance to transitions $\zeta \to \zeta + \delta$ and $\zeta + \delta \to \zeta$. Arrange the passage to the limit so that the time increment δ_t and spatial cell size are such that, in δ_t, a transition of many cells extent takes place and yet the drift parameter varies negligibly. We then have for detailed balance

$$p(\zeta, \zeta_1, \ldots, \zeta_n) \exp\left[-\frac{|\delta - \mu(\zeta; \zeta_1, \ldots, \zeta_n)\delta_t|^2}{2\sigma^2 \delta_t}\right]$$

$$\simeq p(\zeta + \delta, \zeta_1, \ldots, \zeta_n) \exp\left[-\frac{|-\delta - \mu(\zeta; \zeta_1, \ldots, \zeta_n)\delta_t|^2}{2\sigma^2 \delta_t}\right],$$

where for a vector a, $|a|^2 = a^T a$, so that

$$\log p(\zeta + \delta, \zeta_1, \ldots, \zeta_n) - \log p(\zeta, \zeta_1, \ldots, \zeta_n) \simeq \frac{2\mu^T \delta}{\sigma^2}. \quad (6.35)$$

We now revert to the continuous form of the problem and use the Gibbs form of the equilibrium distribution with respect to a Poisson process of unit rate. It follows immediately from (6.35) that with $n + 1$ points in all

$$\mu(\zeta; \zeta_1, \ldots, \zeta_n) \propto \text{grad}_\zeta \phi_{n+1}(\zeta, \zeta_1, \ldots, \zeta_n). \quad (6.36)$$

This shows that if a fixed number of interacting points diffuse so that the process is reversible and has an equilibrium spatial distribution, then the drift parameter is proportional to the gradient of the Gibbs potential. A rigorous proof of this for motion on a torus (Kolmogorov, 1937) was the first result in the theory of spatial–temporal point processes.

If now one combines diffusion with birth and death, separate consideration of the reversibility of various transitions shows that the birth, death and drift rates must be related to the Gibbs potential via (6.33) and (6.36).

For a spatial–temporal process with jump transitions leading to a renewal process on the real line as its equilibrium form (Spitzer, 1977), see Exercise 6.4. An approximate treatment of the temporal properties of the process is sketched in Exercise 6.5.

Bibliographic notes, 6

Spatial point processes were discussed by Matérn (1960) as part of his pioneering study of spatial stochastic processes in general. Bartlett (1975) and Besag (1974) deal with spatial processes on lattices and Ripley (1977) with spatial point processes, all putting some emphasis on statistical analysis.

Haberman (1977) reviews the two fields of mathematical ecology and traffic flow from the viewpoint of classical applied mathematics; point processes arise implicitly rather than explicitly. Pielou (1969) discusses mathematical ecology and Gazis (1974) considers traffic models.

Probabilistic properties of Dirichlet cells associated with a Poisson process are studied by Gilbert (1962) and Miles (1972). Holgate (1972) surveys nearest neighbour distances.

Multidimensional renewal processes were discussed by Bickel and Yahav (1965), Cox (1958), Hunter (1974a, b) and Mode (1967).

The physics of interacting point systems has an extensive and highly specialized literature; see Jeans (1925) for a classical account of kinetic gas theory and Croxton (1974) for a review of recent work on the theory of liquids. An important model for binary variables, in particular spins, on a lattice was introduced by Ising (1925) and has been much studied (McCoy and Wu, 1973).

The pioneering paper on Gibbs distributions as equilibrium distributions of spatial point processes is by Kolmogorov (1937). A general and abstract treatment of Gibbs states is given by Preston (1974, 1976). General theoretical work on spatial–temporal processes has, at the time of writing, concentrated largely on existence and uniqueness results and on the relation between equilibrium spatial processes and Gibbs distributions (Holley and Strook, 1978; Liggett, 1977; Spitzer, 1977). For an abstract treatment of related problems in statistical mechanics, see Ruelle (1969). For the connection with Markov fields, see the account by Ripley and Kelly (1977). Kallenberg (1978) gives some general results about non-interacting systems of points.

Further results and exercises, 6

6.1.   An idealized model of a uniformly exposed and processed photographic film consists of a Poisson process of points in the plane; 'centred' on each point is a random 'transmittance function' and the transmittance at point (ξ, η) in the plane is given by the sum of the separate transmittance functions associated with the individual points. Show that the theory of marked Poisson processes in one dimension can be generalized to obtain, in particular, the second-order properties of the total transmittance function in the plane. [Sections 6.2(i), 5.6(iv); Hamilton, Lawton and Trabka (1972).]

6.2.   Reinterpret the two-dimensional renewal process by taking the two co-ordinate axes to be a common time axis along which two distinct but correlated renewal processes occur. Give the main properties of such a process and explain why it would not often be a reasonable model of, for example, the failure processes for two similar machines. [Sections 6.2(v), 3.2.]

6.3.   A 'hard-core' point process in the plane is produced from a Poisson process of rate ρ by deleting any point within distance r_0 of another point, irrespective of whether that second point has itself already been deleted. Prove that the rate of the derived process is $\rho e^{-\pi r_0^2 \rho}$ and that its maximum as ρ varies is only $\frac{1}{10}$th of the density,

$(\frac{1}{2}r_0^2\sqrt{3})^{-1}$, of a triangular lattice at spacing r_0. Prove also that the conditional intensity is $\rho\exp\{-\pi r_0^2\rho + \rho\gamma(\tau)\}$ for $|\zeta| \ge r_0$ where $\gamma(\zeta)$ is the area common to two circles of radii r_0 and centres distant $|\zeta|$ apart.

A second 'hard-core' model is produced as follows. Points are generated in a Poisson process in space-time at rate ρ. A point is of class U if there has previously occurred a point, of either class, within spatial distance r_0; otherwise the point is of class O. Finally, a spatial process is defined from all those class O points occurring in $0 \le t \le 1$. Prove that the rate of the process is $(1 - e^{-\pi r_0^2\rho})/(\pi r_0^2)$ with a maximum value of $(\pi r_0^2)^{-1}$, and conditional intensity

$$\frac{2\{2\pi r_0^2 - \gamma(\zeta)\} - 2\pi r_0^2(1 - \exp[-\rho\{2\pi r_0^2 - \gamma(\zeta)\}])/(1 - e^{-\pi r_0^2\rho})}{\{2\pi r_0^2 - \gamma(\zeta)\}\{\pi r_0^2 - \gamma(\zeta)\}}.$$

[Section 6.4; Matérn (1960); Paloheimo (1971).]

6.4.   Consider a spatial–temporal process in one spatial dimension, defined as follows. If at time t a point at spatial position ξ has its left and right neighbours at $\xi - u_\xi$ and $\xi + v_\xi$, the point at ξ has probability $\kappa g(u_\xi + v_\xi)\{(g(u_\xi)g(v_\xi)\}^{-1}\delta + o(\delta)$ of changing position in $(t, t + \delta)$, where g is a probability density of a positive random variable and κ is a positive constant. If a transition occurs, the spatial jump has an arbitrary symmetric density. Except for the dependence of jump rate on nearest neighbours, all random variables associated with the process are independent. Suppose that the system starts as an equilibrium renewal process with interval density g. Assuming that this is the equilibrium distribution, verify that the equation of detailed balance is satisfied for transitions in which, denoting spatial intervals by $\{x_r\}$, the point at the left end of the interval x_r jumps to a position inside the interval x_s. To calculate the equilibrium probability density of a particular configuration of points, suppose that the process is defined on a large but finite spatial interval and ignore edge effects. [Section 6.5(v); Spitzer (1977).]

6.5.   Prove that for the spatial-temporal model for a renewal process (Exercise 6.4) the expected jump rate per point is $\kappa\mu_g$, where μ_g is the mean of the density g. Suppose that it is possible to find a spatial interval (or set) I large enough to contain quite a few points and yet such that any point in I experiencing a jump is virtually certain to leave I. By noting that for sufficiently small t, $\kappa\mu_g t$ is the probability that an arbitrary point jumps in time t, show that the correlation between the numbers of points in I a time t apart is approximately

$$1 - 2\kappa\mu_g t + \kappa\mu_g t \frac{E\{N(I)\}}{\text{var}\{N(I)\}} + o(t),$$

where $N(I)$ is the number of points in I at an arbitrary time.

When the renewal process is a Poisson process, show that this is $1 - \kappa\mu_g t + o(t)$. Enlarge on the difficulty of obtaining more detailed results for this correlation. [Section 6.5(v).]

6.6.   Review the special one-dimensional models of Chapter 3 to determine which can conveniently be regarded as equilibrium distributions in space for a suitable reversible spatial–temporal model. [Section 6.5(v).]

References

Aalen, O. (1978). Nonparametric inference for a family of counting processes. *Ann. Statist.*, **6**, 701–26.

Aalen, O. O. and Hoem, J. M. (1978). Random time changes for multivariate counting processes. *Scand. Actuarial J.*, 81–101.

Bartlett, M. S. (1955). Discussion of paper by D. R. Cox. *J. R. Statist. Soc.*, B **17**, 159–60.

Bartlett, M. S. (1963). The spectral analysis of point processes (with discussion). *J. R. Statist. Soc.*, B **25**, 264–96.

Bartlett, M. S. (1964). The spectral analysis of two-dimensional point processes. *Biometrika* **51**, 299–311.

Bartlett, M. S. (1975). *The Statistical Analysis of Spatial Pattern*. London: Chapman and Hall.

Bartlett, M. S. (1978). *An Introduction to Stochastic Processes*. 3rd Edn. Cambridge University Press.

Berman, M. (1977). Some multivariate generalizations of results in univariate stationary point processes. *J. Appl. Prob.*, **14**, 748–57.

Berman, M. (1978). Regenerative multivariate point processes. *Adv. Appl. Prob.*, **10**, 411–30.

Besag, J. (1974). Spatial interaction and the statistical analysis of lattice systems (with discussion). *J. R. Statist. Soc.*, B **36**, 192–236.

Bhabha, H. J., and Heitler, W. (1937). The passage of fast electrons and the theory of cosmic showers. *Proc. Roy. Soc.*, A **159**, 432–58.

Bickel, P. J. and Yahav, J. A. (1965). Renewal theory in the plane. *Ann. Math. Statist.*, **36**, 946–55.

Boel, R., Varaiya, P. and Wong, E. (1975). Martingales on jump processes. I and II. *S.I.A.M. J. Control* **13**, 999–1061.

Brémaud, P. and Jacod, J. (1977). Processus ponctuels et martingales: résultats récents sur la modélisation et le filtrage. *Adv. Appl. Prob.*, **9**, 362–416.

Brillinger, D. R. (1975). The identification of point process systems (with discussion). *Ann. Prob.*, **3**, 909–29.

Brillinger, D. R. (1978a). Comparative aspects of the study of ordinary time series and of point processes, in *Developments in Statistics*, Vol. I, ed. P. R. Krishnaiah, pp. 33–133. New York: Academic Press.

Brillinger, D. R. (1978b). A note on a representation for the Gauss–Poisson process. *Stoch. Proc. Appl.*, **6**, 135–7.

Chiang, C. L. (1968). *Introduction to Stochastic Processes in Biostatistics*. New York: Wiley.

Çinlar, E. (1969). Markov renewal theory. *Adv. Appl. Prob.*, **1**, 123–87.

Çinlar, E. (1972). Superposition of point processes, in *Stochastic Point Processes*, ed. P. A. W. Lewis, pp. 549–606. New York: Wiley.

Çinlar, E. and Jagers, P. (1973). Two mean values which characterize the Poisson process. *J. Appl. Prob.*, **10**, 678–81.

Clausius, R. (1858). Ueber die Mittlere Länge der Wege...," *Annalen der Physik* **105**, 239–58. Reprinted in English translation, (1969), *Kinetic Theory*, Vol. 1, ed. S. G. Brush, pp. 135–47. Oxford: Pergamon.

Cox, D. R. (1955). Some statistical methods connected with series of events (with discussion). *J. R. Statist. Soc.*, B **17**, 129–64.

Cox, D. R. (1958). Discussion of paper by W. L. Smith. *J. R. Statist. Soc.*, B **20**, 286–7.

Cox, D. R. (1962). *Renewal Theory*. London: Methuen.

Cox, D. R. (1963). Some models for series of events. *Bull. I.S.I.* **40**, 737–46.

Cox, D. R. (1969). Some sampling problems in technology, in *New Developments in Survey Sampling*, eds. N. L. Johnson and H. Smith, pp. 506–27. New York: Wiley.

Cox, D. R. and Isham, V. (1977). A bivariate point process connected with electronic counters. *Proc. Roy. Soc.*, A **356**, 149–60.

Cox, D. R. and Isham, V. (1978). Series expansions for the properties of a birth process of controlled variability. *J. Appl. Prob.*, **15**, 610–16.

Cox, D. R. and Lewis, P. A. W. (1966). *The Statistical Analysis of Series of Events*. London: Methuen.

Cox, D. R. and Lewis, P. A. W. (1972). Multivariate point processes. *Proc. 6th Berkeley Symp.*, **3**, 401–48.

Cox, D. R. and Miller, H. D. (1965). *The Theory of Stochastic Processes*. London: Methuen.

Cox, D. R. and Smith, W. L. (1953). The superposition of several strictly periodic sequences of events. *Biometrika* **40**, 1–11.

Cox, D. R. and Smith, W. L. (1954). On the superposition of renewal processes. *Biometrika* **41**, 91–9.

Cramér, H. and Leadbetter, M. R. (1967). *Stationary and Related Stochastic Processes*. New York: Wiley.

Croxton, C. A. (1974). *Liquid State Physics – a Statistical Mechanical Introduction*. Cambridge University Press.

Daley, D. J. (1974). Various concepts of orderliness for point-processes, in *Stochastic Geometry*, eds. E. F. Harding and D. G. Kendall, pp. 148–61. London: Wiley.

Daley, D. J. and Milne, R. K. (1973). The theory of point processes: a bibliography. *Rev. I.S.I.* **41**, 183–201.

Daley, D. J. and Milne, R. K. (1975). Orderliness, intensities and Palm–Khinchin equations for multivariate point processes. *J. Appl. Prob.*, **12**, 383–9.

Daley, D. J. and Vere–Jones, D. (1972). A summary of the theory of point processes, in *Stochastic Point Processes*, ed. P. A. W. Lewis, pp. 299–383. New York: Wiley.

Davis, M. H. A. (1976). The representation of martingales of jump processes. *S.I.A.M.J. Control* **14**, 623–38.

Doob, J. L. (1953). *Stochastic Processes*. New York: Wiley.

Ekholm, A. (1972). A generalization of the two-state two-interval semi-Markov model, in *Stochastic Point Processes*, ed. P. A. W. Lewis, pp. 272–84. New York: Wiley.

Erickson, K. B. and Guess, H. (1973). A characterization of the exponential law. *Ann. Prob.*, **1**, 183–5.

Feller, W. (1968, 1971). *Introduction to Probability Theory and its Applications*, Vol. 1, 3rd Edn. Vol. 2, 2nd Edn. New York: Wiley.

Fienberg, S. (1974). Stochastic models for single neutron firing trains: a survey. *Biometrics* **30**, 399–427.

Gaver, D. P. (1963). Random hazard in reliability problems. *Technometrics* **5**, 211–26.

Gaver, D. P. (1976). Random record models. *J. Appl. Prob.*, **13**, 538–47.

Gazis, D. C. (ed.) (1974). *Traffic Science*. New York: Wiley.

Gilbert, E. N. (1962). Random subdivisions of space into crystals. *Ann. Math. Statist.*, **33**, 958–72.

Gnedenko, B. V. and Kovalenko, I. N. (1968). *Introduction to Queueing Theory*. Jerusalem: Israel Program for Scientific Translations.

Goldman, J. R. (1967). Stochastic point processes: limit theorems. *Ann. Math. Statist.*, **38**, 771–9.

Govier, L. J. and Lewis, T. (1961). Serially correlated arrivals in some queueing and inventory systems. *Proc. 2nd Int. Conf. on Operational Research*, 355–64.

Govier, L. J. and Lewis, T. (1963). Stock levels generated by a controlled-variability arrival process. *Operat. Res.*, **11**, 693–701.

Grandell, J. (1976). *Doubly stochastic Poisson processes. Lecture notes in mathematics*, **529**. Berlin: Springer–Verlag.

Greenwood, M. and Yule, G. U. (1920). An inquiry into the nature of frequency distributions representative of multiple happenings with particular reference to the occurrence of multiple attacks of disease or of repeated accidents. *J. R. Statist. Soc.*, **83**, 255–79.

Grigelionis, B. (1963). On the convergence of sums of random step processes to a Poisson process. *Theory Prob. Appl.*, **8**, 177–82.

Haberman, R. (1977). *Mathematical Models*. Englewood Cliffs: Prentice-Hall.

Hamilton, J. F., Lawton, W. H. and Trabka, E. A. (1972). Some spatial and temporal point processes in photographic science, in *Stochastic Point Processes*, ed. P. A. W. Lewis, pp. 817–67. New York: Wiley.

Harris, T. E. (1971). Random motions and point processes. *Z. Wahrschein.*, **18**, 85–115.

Hawkes, A. G. (1971a). Spectra of some self-exciting and mutually exciting point processes. *Biometrika* **58**, 83–90.

Hawkes, A. G. (1971b). Point spectra of some mutually exciting point processes. *J. R. Statist. Soc.*, B **33**, 438–43.

Hawkes, A. G. (1972). Spectra of some mutually exciting point processes with associated variables, in *Stochastic Point Processes*, ed. P. A. W. Lewis, pp. 261–71. New York: Wiley.

Hawkes, A. G. and Oakes, D. (1974). A cluster process representation of a self-exciting process. *J. Appl. Prob.*, **11**, 493–503.

Holgate, P. (1972). The use of distance methods for the analysis of spatial distribution of points, in *Stochastic Point Processes*, ed. P. A. W. Lewis, pp. 122–35. New York: Wiley.

Holley, R. A. and Strook, D. W. (1978). Nearest neighbor birth and death processes on the real line. *Acta Math.* **140**, 103–54.

Hunter, J. J. (1974a). Renewal theory in two dimensions: basic results. *Adv. Appl. Prob.*, **6**, 376–91.

Hunter, J. J. (1974b). Renewal theory in two dimensions: asymptotic results. *Adv. Appl. Prob.* **6**, 546–62.

Isham, V. (1975). On a point process with independent locations. *J. Appl. Prob.*, **12**, 435–46.

Isham, V. (1977a). Constructions for planar point processes using concentric circles. *Stoch. Proc. Appl.*, **5**, 131–41.

Isham, V. (1977b). A Markov construction for a multidimensional point process. *J. Appl. Prob.*, **14**, 507–15.

Isham, V. (1980). Dependent thinning of point processes. *J. Appl. Prob.*, **17**, to be published.

Isham, V., Shanbhag, D. N. and Westcott, M. (1975). A characterization of the Poisson process using forward recurrence times. *Math. Proc. Camb. Phil. Soc.* **78**, 513–16.

Isham, V. and Westcott, M. (1979). A self-correcting point process. *Stoch. Proc. Appl.*, **8**, 335–48.

Ising, E. (1925). Beitrag Zur Theorie des Ferromagnetismus. *Z. Physik.*, **31**, 253–8.

Jacobs, P. A. and Lewis, P. A. W. (1977). A mixed autoregressive-moving average exponential sequence and point process (EARMA 1, 1). *Adv. Appl. Prob.*, **9**, 87–104.

Jacod, J. (1975). Multivariate point process: predictable projection. Radon–Nikodym derivatives, representation of martingales. *Z. Wahrschein.*, **31**, 235–53.

James, G. S. (1952). Notes on a theorem of Cochran. *Proc. Camb. Phil. Soc.*, **48**, 443–6.

Jarrett, R. G. (1979). A note on the intervals between coal-mining disasters. *Biometrika* **66**, 191–3.

Jeans, J. H. (1925). *The Dynamical Theory of Gases*, 4th Edn. Cambridge University Press.

Jensen, A. (1948). An elucidation of Erlang's statistical works through the theory of stochastic processes. *The Life and Works of A. K. Erlang*, pp. 23–100, Copenhagen Tel. Co.

Johnson, N. L. and Kotz, S. (1969). *Distributions in Statistics: Discrete Distributions.* Boston: Houghton–Mifflin.

Kallenberg, O. (1973). Characterization and convergence of random measures and point processes. *Z. Wahrschein.,* **27**, 9–21.

Kallenberg, O. (1975). Limits of compound and thinned point processes. *J. Appl. Prob.,* **12**, 269–78.

Kallenberg, O. (1976). *Random Measures.* Berlin: Akademie Verlag.

Kallenberg, O. (1978). On the independence of velocities in a system of noninteracting particles. *Ann. Prob.,* **6**, 885–90.

Kendall, D. G. (1942). A summation formula associated with finite trigonometric integrals *Quart. J. Maths.,* **13**, 172–84.

Kerstan, J. and Matthes, K. (1964). Stationäre zufällige Punktfolgen, II. *Jber. Deutsch. Math.-Verein.,* **66**, 106–18.

Khintchine, A. Y. (1960). *Mathematical Methods in the Theory of Queueing.* London: Griffin.

Kingman, J. F. C. (1964). On doubly stochastic Poisson processes. *Proc. Camb. Phil. Soc.,* **60**, 923–30.

Kingman, J. F. C. (1977). The asymptotic covariance of two counters. *Math. Proc. Camb. Phil. Soc.,* **82**, 447–52.

Kittel, C. (1958). *Elementary Statistical Physics.* New York: Wiley.

Kolmogorov, A. (1937). Zur Umkehrbarkeit der Statistichen Naturgesetze. *Math. Ann.,* **113**, 766–72.

Krickeberg, K. (1974). Moments of point-processes, in *Stochastic Geometry,* eds. E. F. Harding and D. G. Kendall, pp. 89–113. London: Wiley.

Lai, C. D. (1978). An example of Wold's point processes with Markov-dependent intervals. *J. Appl. Prob.,* **15**, 748–58.

Lawrance, A. J. (1972). Some models for stationary series of univariate events, in *Stochastic Point Processes,* ed. P. A. W. Lewis, pp. 199–256. New York: Wiley.

Lawrance, A. J. (1973). Dependency of intervals between events in superposition processes. *J. R. Statist. Soc.,* B **35**, 306–15.

Lawrance, A. J. and Lewis, P. A. W. (1977). An exponential moving-average sequence and point process (EMA1). *J. Appl. Prob.,* **14**, 98–113.

Leadbetter, M. R. (1968). On three basic results in the theory of stationary point processes. *Proc. Amer. Math. Soc.,* **19**, 115–7.

Leadbetter, M. R. (1972a). On basic results of point process theory. *Proc. 6th Berkeley Symp.,* **3**, 449–62.

Leadbetter, M. R. (1972b). Point processes generated by level crossings, in *Stochastic Point Processes,* ed. P. A. W. Lewis, pp. 436–67. New York: Wiley.

Lee, P. M. (1967). Infinitely divisible stochastic processes. *Z. Wahrschein.,* **7**, 147–60.

Lévy, P. (1956). Processus semi-Markoviens. *Proc. Int. Cong. Math. 1954 (Amsterdam)* **3**, 416–26.

Lewis, P. A. W. (1964). A branching Poisson process model for the analysis of computer failure patterns (with discussion). *J. R. Statist. Soc.*, B **26**, 398–456.

Lewis, P. A. W. (ed.) (1972). *Stochastic Point Processes*. New York: Wiley.

Lewis, T. (1961). The intervals between regular events displaced in time by independent random deviations of large dispersion. *J. R. Statist. Soc.*, B **23**, 476–83.

Lewis, T. and Govier, L. J. (1964). Some properties of counts of events for certain types of point process. *J. R. Statist. Soc.*, B **26**, 325–37.

Liggett, T. M. (1977). *The stochastic evolution of infinite systems of interacting particles. Lecture notes in mathematics*, **598**, pp. 187–248. Berlin: Springer–Verlag.

Liptser, R. S. and Shiryayev, A. N. (1977, 1978). *Statistics of Random Processes*, Vols. 1 and 2. New York: Springer–Verlag.

Macchi, O. (1979). Stochastic point processes in pure and applied physics. *Bull. I.S.I.* **47**, to be published.

Mandelbrot, B. B. (1977). *Fractals: Form, Chance and Dimension*. San Francisco: Freeman.

Matérn, B. (1960). Spatial variation. *Meddelanden från Statens Skogsforskningsinstitut* **49**, nr. 5.

Matthes, K. (1964). Stationäre zufällige Punktfolgen, I. *Jber. Deutsch. Math.–Verein.*, **66**, 66–79.

Matthes, K. (1972). Infinitely divisible point processes, in *Stochastic Point Processes* ed. P. A. W. Lewis, pp. 384–404. New York: Wiley.

Matthes, K., Kerstan, J. and Mecke, J. (1978). *Infinitely Divisible Point Processes*. London: Wiley.

McCoy, B. M. and Wu, Tai Tsun (1973). *The Two-Dimensional Ising Model*. Cambridge, Mass: Harvard University Press.

McFadden, J. A. (1962). On the lengths of intervals in stationary point processes. *J. R. Statist. Soc.*, B, **24**, 364–82.

Miles, R. E. (1972). The random division of space (with discussion). *Adv. Appl. Prob. Suppl.*, 243–66. (Proc. Symp. Statistical and Probabilistic Problems in Metallurgy.)

Milne, R. K. (1974). Infinitely divisible bivariate Poisson processes. *Adv. Appl. Prob.*, **6**, 226–7.

Milne, R. K. and Westcott, M. (1972). Further results for Gauss–Poisson processes. *Adv. Appl. Prob.*, **4**, 151–76.

Mode, C. J. (1967). A renewal density theorem in the multidimensional case. *J. Appl. Prob.*, **4**, 62–76.

Moran, P. A. P. (1950). Numerical integration by systematic sampling. *Proc. Camb. Phil. Soc.*, **46**, 111–5.

Moran, P. A. P. (1967). A non-Markovian quasi-Poisson process. *Studia Sci. Math. Hungar.*, **2**, 425–9.

Moyal, J. E. (1962). The general theory of stochastic population processes. *Acta Math.*, **108**, 1–31.

Müller, J. W. (ed.) (1975). *Bibliography on Dead-time Effects*. Bureau international des Poids et Mesures. Sèvres, France, Report 75/6.

Naus, J. I. (1966). Some probabilities, expectations and variances for the size of largest clusters and smallest intervals. *J. Amer. Statist. Assoc.*, **61**, 1191–9.

Naus, J. I. (1979). An indexed bibliography of clusters, clumps and coincidences. *Rev. I.S.I.* **47**, 47–78.

Newman, D. S. (1970). A new family of point processes which are characterized by their second moment properties. *J. Appl. Prob.*, **7**, 338–58.

Neyman, J. and Scott, E. L. (1958). A statistical approach to problems of cosmology (with discussion). *J. R. Statist. Soc.*, B **20**, 1–43.

Neyman, J. and Scott, E. L. (1972). Processes of clustering and applications, in *Stochastic Point Processes*, ed. P. A. W. Lewis, pp. 646–81. New York: Wiley.

Oakes, D. (1972). *Semi-Markov representations of some stochastic point processes.* Ph.D. thesis, University of London.

Oakes, D. (1975). The Markovian self-exciting process. *J. Appl. Prob.*, **12**, 69–77.

Oakes, D. (1976). Bivariate Markov processes of intervals. *Information and Control* **32**, 231–41.

Palm, C. (1943). Intensitätsschwankungen im Fernsprechverkehr. *Ericsson Techniks*, No. 44.

Paloheimo, J. E. (1971). On a theory of search. *Biometrika* **58**, 61–75.

Papangelou, F. (1972). Integrability of expected increments of point processes and a related random change of scale. *Trans. Amer. Math. Soc.*, **165**, 483–506.

Papangelou, F. (1974). The conditional intensity of general point processes and an application to line processes. *Z. Wahrschein.*, **28**, 207–26.

Pielou, E. C. (1969). *Introduction to mathematical ecology*. New York: Wiley.

Preston, C. J. (1974). *Gibbs States on Countable Sets*. Cambridge University Press.

Preston, C. J. (1976). *Random Fields, Lecture notes in Mathematics*, **534**, Berlin: Springer–Verlag.

Pyke, R. (1961a). Markov renewal processes: definitions and preliminary properties. *Ann. Math. Statist.*, **32**, 1231–42.

Pyke, R. (1961b). Markov renewal processes with finitely many states. *Ann. Math. Statist.*, **32**, 1243–59.

Pyke, R. and Schaufele, R. (1964). Limit theorems for Markov renewal processes. *Ann. Math. Statist.*, **35**, 1746–64.

Pyke, R. and Schaufele, R. (1966). The existence and uniqueness of stationary measures for Markov renewal process. *Ann. Math. Statist.*, **37**, 1439–62.

Rényi, A. (1956). A characterization of Poisson processes. *Magyar Tud. Akad. Mat. Kutato Int. Közl.*, **1**, 519–27. Translated in *Selected Papers of Alfréd Rényi*, Vol. 1, ed. Pál Turán, pp. 622–8. Akadémiai Kiadó: Budapest, 1976.

Rice, S. O. (1944). Mathematical analysis of random noise. *Bell Syst. Tech. J.*, **23**, 282–332.

Rice, S. O. (1945). Mathematical analysis of random noise. *Bell. Syst. Tech. J.*, **24**, 46–156 [previous reference continued].

Ripley, B. D. (1977). Modelling spatial patterns (with discussion). *J. R. Statist. Soc.*, B **39**, 172–212.

Ripley, B. D. and Kelly F. P. (1977). Markov point processes. *J. London Math. Soc.* (2), **15**, 188–92.

Rudemo, M. (1973). On a random transformation of a point process to a Poisson process, in *Mathematics and Statistics: essays in honour of Harald Bergström*, eds. P. Jagers and L. Rade, pp. 79–85. Göteborg.

Ruelle, D. (1969). *Statistical Mechanics*. New York: Benjamin.

Ryll–Nardzewski, C. (1961). Remarks on processes of calls. *Proc. 4th Berkeley Symp.*, **2**, 455–65.

Sampath, G. and Srinivasan, S. K. (1977). *Stochastic models for spike trains of single neurons. Lecture notes in biomathematics*, **16**. Berlin: Springer–Verlag.

Samuels, S. M. (1974). A characterization of the Poisson process. *J. Appl. Prob.*, **11**, 72–85.

Simon, B. (1974). *The $P(\phi)_2$ Euclidean (Quantum) Field Theory*. Princeton University Press.

Sinai, Ya. (1963). On the foundations of the ergodic hypothesis for a dynamical system of statistical mechanics. *Sov. Math. Dokl.*, **4**, 1818–22.

Smith, W. L. (1955). Regenerative stochastic processes. *Proc. Roy. Soc.*, A **232**, 6–31.

Smith, W. L. (1958). Renewal theory and its ramifications (with discussion). *J. R. Statist. Soc.* B **20**, 284–302.

Snyder, D. L. (1975). *Random Point Processes*. New York: Wiley.

Spitzer, F. (1977). Stochastic time evolution of one dimensional infinite particle systems. *Bull. Amer. Math. Soc.*, **83**, 880–90.

Srinivasan, S. K. (1974). *Stochastic Point Processes*. London: Griffin.

Srinivasan, S. K. (1969). *Stochastic Theory and Cascade Processes*. New York: American Elsevier.

Stone, C. (1968). On a theorem of Dobrushin. *Ann. Math. Statist.*, **39**, 1391–401.

Tabor, D. (1969). *Gases, Liquids and Solids*. Harmondsworth: Penguin.

Vere–Jones, D. (1970). Stochastic models for earthquake occurrence (with discussion). *J. R. Statist. Soc.*, B **32**, 1–62.

Vit, P. (1974). On the equivalence of certain truncated point processes. *J. Appl. Prob.*, **11**, 601–4.

Warren, W. G. (1971). The centre-satellite concept as a basis for ecological sampling (with discussion), in *Statistical Ecology*, Vol. 2, eds. G. P. Patil, E. C. Pielou and W. E. Waters, pp. 87–118. University Park: Penn. State University Press.

Weiss, G. (1973). *Filtered Poisson processes as models for daily streamflow data*. Ph.D. thesis, University of London.

Westcott, M. (1972). The probability generating functional. *J. Austral. Math. Soc.*, **14**, 448–66.

Westcott, M. (1976). Simple proof of a result on thinned point processes. *Ann. Prob.* **4**, 89–90.

Westcott, M. (1977). The random record model. *Proc. Roy. Soc.*, A **356**, 529–47.

Whittle, P. (1975). Reversibility and acyclicity, in *Perspectives in Probability and Statistics*, ed. J. Gani, pp. 217–24. London: Academic Press.

Wisniewski, T. K. M. (1972). Bivariate stationary point processes, fundamental relations and first recurrence times. *Adv. Appl. Prob.*, **4**, 296–317.

Wold, H. (1948). On stationary point processes and Markov chains. *Skand. Aktuar.*, **31**, 229–40.

Wold, H. (1949). Sur les processus stationnaires ponctuels. *Colloques Internationaux du CNRS* **13**, 75–86.

Zitek, F. (1957). On a theorem of Korolyuk. *Czech. Math. J.*, **7**, 318–19.

Author index

Subject index